THE GREAT RIFT

THE
GREAT RIFT
Africa's Changing Valley

ANTHONY SMITH

Sterling Publishing Co., Inc.   New York

For Joan

ACKNOWLEDGEMENTS

As *The Great Rift* is partner to both a television
and a radio series my first thanks are to the BBC individuals
responsible for those programmes whose assistance was
also invaluable in the creation of this book, notably
John Sparks, Andrew Neal, Adrian Warren, Keith Scholey,
Michael Bright and John Harrison.
Many other people helped, by being interviewed
or assisting with the manuscript, and extra thanks are
therefore due to: Frank Barnes, Graham Berry, Joseph Boit,
Pierre Brichard, Martin Clarke, Harvey Croze, Paul Dickson,
Robert Foley, Lynne Frostick, Humphry Greenwood,
Osker Grutten, Michael Gwynne, John Harris, Chris Hillman,
Michael Jennings, Hugh Lamprey, Mary Leakey, Rosemary
Lowe-McConnell, Walter Lusigi, Elizabeth Meyerhoff-Roberts,
Godfrey Ole Moita, Celia Nyamweru, Titus Odembo,
Simon Peel, Mike Reynolds, Joan Root, Denis Stiles, Mathew
Ole Suya, Peter Vine, Maurice Wambua, Geoff and Hilary
Welch, Nassus Yarinakis. Without such considerable
cooperation the writing of a book such as this would not
merely be much harder; it would be quite impossible.

A.S.

Copyright © 1988 by Anthony Smith
Published in 1989 by Sterling Publishing Co., Inc.,
387 Park Avenue South, New York, N.Y. 10016
Distributed in Canada by Oak Tree Press Ltd
c/o Canadian Manda Group, P.O. Box 920, Station U
Toronto, Ontario, Canada M8Z 5P9
First published in the United Kingdom in 1988 by
BBC Books a division of BBC Enterprises Limited,
London, England.
Sterling ISBN 0–8069–6906–7
Printed in Great Britain

89-1468

CONTENTS

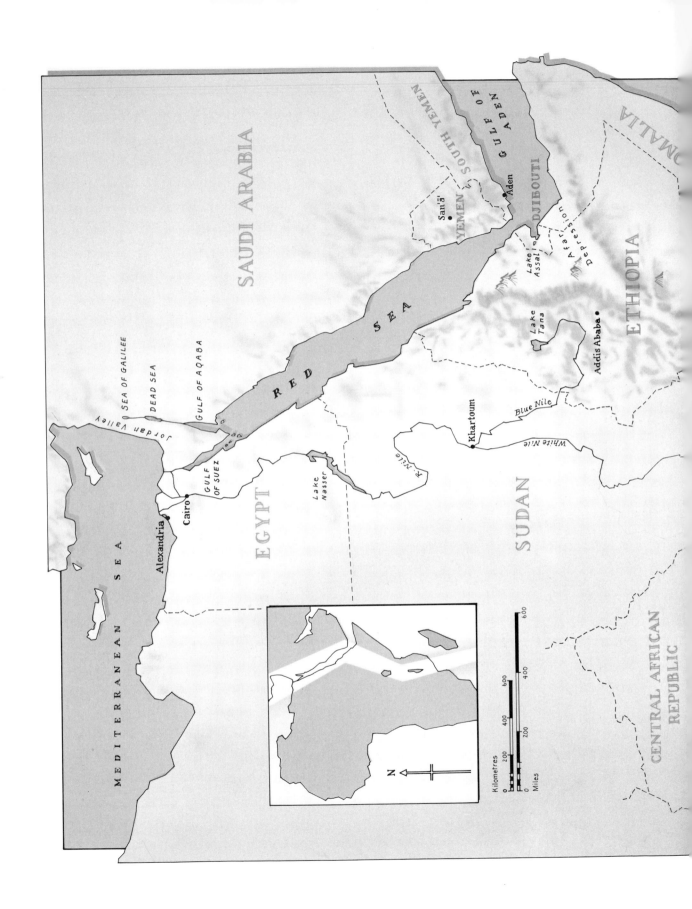

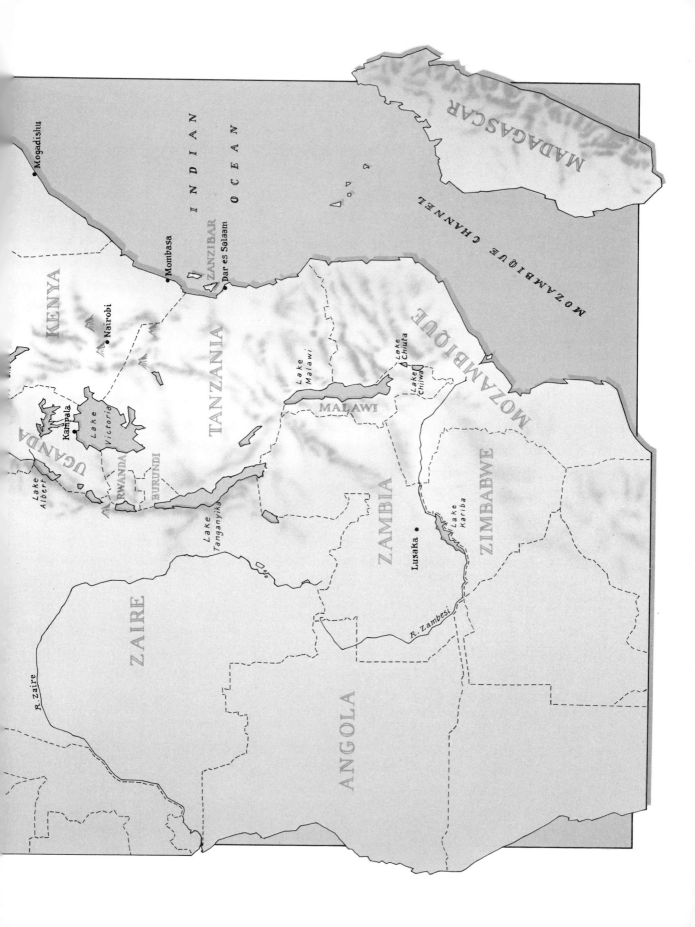

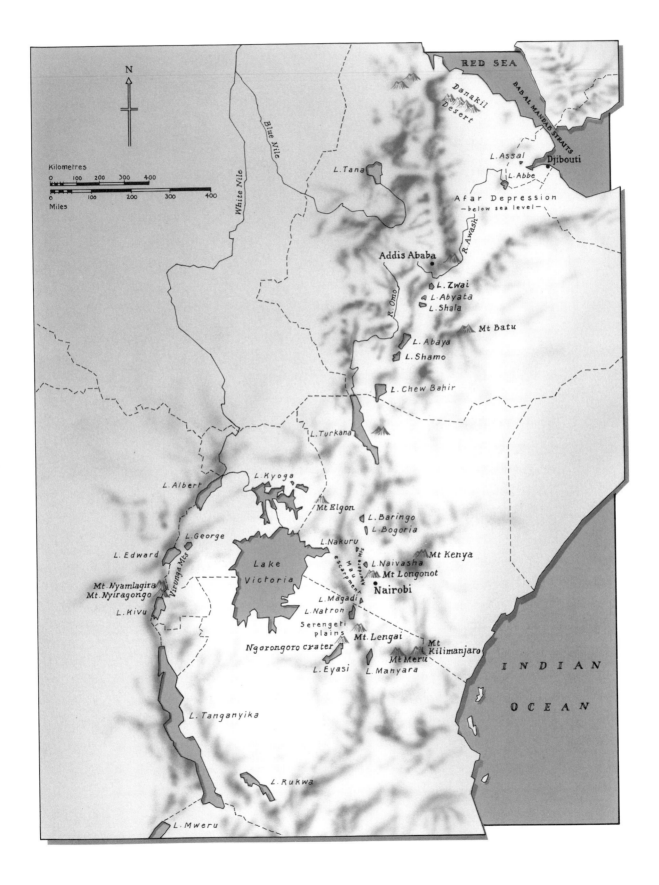

INTRODUCTION

—

The view of the Rift made a tremendous impression, partly because I was terrified. To travel by balloon should be a form of perfection, but this journey was not going according to plan, except for the astonishing spectacle of the planet's outer crust which it afforded. The land was like none other I had ever seen, and its nature only added to my fear. Mountains often have cliffs, but not, in general, a succession of steep descents. The ground fell away dramatically, as if great steps had been carved in the rock and, as the three of us travelled, each huge step down meant that our vantage point became ever further above the ground. From a maximum of 2700 metres above sea level the land in that area jerkily dropped to 600 metres; and, as we progressed across this steep descent, our wicker basket became an increasingly frightening place in which to be.

The departure from Nairobi had been a hasty affair. We had planned to ascend at midday, when the area's atmospheric contortions would be less stimulating, but our hydrogen cylinders proved to be insufficiently charged with gas and finding more was not an easy task at a weekend. Inflating the balloon became steadily more awkward, as increasing gustiness threw us this way and that. Eventually, and belatedly, without a proper appreciation of either the wind's strength or the thermals' vigour, we set off. As instructed, the many helpers on the ground let go their grip, and another form of grip immediately took charge. We were hurtled over the Nairobi airfield, an army camp, suburban homes, and then the Athi plains. It was all so effortless, and alarming.

A gentle wind considers the objects in its path, anticipating them and riding past or over them. The wind that had us in its power was not behaving gently. It rushed along with as much concern for other things as an avalanche. We saw ahead, with increasing clarity, the Ngong hills. They always look well, but they appeared infinitely more dramatic as we approached headlong at some 65 kilometres an hour. Our wind should have carried us over them, like water arching its back over a boulder in the stream. Instead, it became increasingly obvious that we were to hit them amidships. Our view narrowed to a farmer standing by a patch of corn at the side of his hut. He ought to have been impressed as 20 metres of orange and silver balloon came hurtling towards his field, but he stared fixedly at us, giving no sign of apprehension. We shouted some greeting, hit, winced, rasped momentarily through his corn, and then were on our way again, now with an upward speed to match that horizontal haste.

Within moments the Ngongs had resumed their normal, long-range aspect, and the magnificent view presented itself – the step by step descents, the faults and plateaus by which the Ngongs drop down to the level of Magadi, some 50 kilometres ahead. As we climbed, in the biggest bounce any of us had known, we were soon a kilometre above that solitary farmer. Balloons should not bounce, at least not with such gargantuan leaps, but the wind was plainly cavorting over those Ngongs and we were merely one more piece of it. Like a cork on water we had no control over our destination. Once we reached our maximum height, we would plummet downwards from our eagle's perch, and hit the ground again. There was no point in releasing gas, and the ballast on board was quite inadequate to curtail this form of high jinks. Our only hope was to pull the rip and release all the gas just before contact with the ground.

That first bounce lasted about twenty minutes. As we slowly soared high above the ground, and waited there for gravity to take hold, we had ample occasion to absorb the extraordinary scenery. Then suddenly we were descending with increasing verve. When we saw the piece of ground poised to welcome us, any hope of pulling the rip was dashed. Awaiting us was a medley of tremendous rocks perched on the summit of another range of hills. We would hit, we would certainly be knocked savagely against each other, and then we would be lifted once more into the relative calm of another ascent. And that, more or less, is how it happened.

An immediate post-mortem was impossible as we rose again in altitude, and that astonishing view of ridges and faults, of staggered drops in the planet's surface, once more seared itself into all our memories. This time we were permitted an even more panoramic picture as the balloon soared even higher above the ground. The steps of land appeared a touch flattened but, by looking north and south, we could observe how they dropped down to form a great valley before climbing again on its other side. The white smear of Lake Magadi, glimmering from the soda of which it is composed, lay at the base of both inclines. So too did Lake Natron, with an extra fuzz of pink from all its flamingoes. There were few clouds down below, but it was the contorted land that most astonished us. The Earth's crust is usually hidden beneath soil and crops or by distracting objects made by people. On that day, and from our pin-prick in the sky, we could see it all – how it must have been made, which bits were tugging apart, where weaknesses had been formed. For once geomorphology had become a subject for anyone with eyes.

Our wonder at the scene became interwoven with apprehension about the landing ahead. We knew we would hit; but would the encounter be with a tree or a patch of dusty ground? As we descended the possibilities narrowed to a plateau, a jagged fault, or a second flatter piece progressing not too far ahead. It seemed at first as if it might be the awful cliff, but then it became the piece of flatter ground. Rocks sped past under the balloon and then we hit a tree violently. It cracked, or we did, and we quickly pulled at the rip. The balloon leapt up, there was another crack as we collided with a second tree and thorns rasped the basket's outer side. Then we gently touched the earth, and the flapping balloon lay down and died, its gas rushing skywards from the open slit of a rip. We laughed, and moved from the major worry of a frightful death to a myriad minor concerns, such as a bleeding arm, a collected crown of thorns, and a pain here or there which seemed to have no cause. After we had scrambled out, there were worries even more trivial, such as where exactly we were.

Finally came the realisation that in no other manner could we have experienced the Rift Valley so vividly. Not only had a large part of it been spread out below us, but its layering of contours had been entirely relevant to our flight's progress. From the Ngong hills to the Magadi plain, we had gained in altitude at the apex of each bounce. When we finally came down to earth, we were much lower than at our starting point. It was very much hotter and drier, with a different kind of thorn. We were at the bottom of the Rift, that spectacular cleavage in the planet's surface that always astonishes, but nowhere more so than in Kenya. We had observed one of the greatest sights on Earth. More amazing still, we had lived to tell the tale.

No one can be unmoved by their first experience of the Rift Valley, however and wherever that happens to be. Seen as a whole, the Rift runs southwards from Turkey, through Israel, along the Gulf of Aqaba, down the Red Sea, across the highlands of Ethiopia, through Kenya and its lakes, down through the middle of Tanzania and into Malawi. There is a secondary or Western Rift travelling along the borderland between Zaire and its eastern neighbours, Uganda, Rwanda, Burundi and Tanzania. Practically every lake in eastern Africa lies within some section of the Rift, the two important exceptions being the feeders of the Nile: Lake Tana, in western Ethiopia, which starts the Blue Nile with a flourish, and Lake Victoria, shared by Uganda, Kenya and Tanzania, which serves the same reservoir role for the White Nile. However, the Rift Valley is not just a single or a double crack. There are faults, ridges, slips, overlays, upthrusts, and then all manner of volcanic events forming a widespread Rift Valley *system*.

My first view of it is more embedded in the past than my aerial encounter with the Ngongs and can serve to introduce the Rift as it was introduced to me. The fact of knowing nothing about Africa was once the principal spur for buying a motorbike. Having finished a job in Cape Town, I was presented with a steamer ticket back to London. With all the impudence of youth I cashed the reservation to provide the where-withal for 200 ccs of Triumph Tiger Cub. Two wheels seemed to provide a better platform for returning to England than the tedium of a ship's deck, and the bike would permit me to experience the length, if not the breadth, of Africa.

Personal ignorance of the continent as I left Cape Town was con-siderable, and my lack of preparation concerning the land ahead of me had perhaps been carried a touch too far. I quickly became lost, not having planned a route, and was then blinded by two flies working in tandem. Much time passed before I could see, let alone find, a shop with goggles for sale. Then, having ascended the 900 metres of the Hex River

Pass, I froze. This is supposed to be Africa, I chattered to myself, as I put on extra shirts while the hail bounced ahead of me. At Laingsburg, 290 body-stiffening kilometres from Cape Town, I managed to equip myself with gauntlets, lots of wool, and a warm coat. I still knew nothing of the many thousand kilometres that lay ahead, but inside those layers, and behind those goggles, I felt better prepared for whatever the continent had in store.

For the next few weeks the main impression was one of huge distances. I had never known such space. I would labour all morning up some incline, and finally reach its gently arching summit to be rewarded with nothing more than a mirror image on the other side. On one occasion, being curious about my beetle-like progression through this great quantity of world, I stopped the bike and left its engine running while the machine stood on its rest. I walked away from it, up a slope, and decided to keep on walking until I could hear the thing no more. Eventually, confused between the local grasshoppers and that distant motorbike, I stopped and turned around. The sight of my microscopic machine and, seemingly, all of Africa put my journey into harsh perspective. We were a minute event, set within a tremendous context.

Another reason for this diversion was that I was finding Africa a touch uniform. After the climb from Cape Town, and all the exciting kopjes of the Karroo, the continent seemed to settle into a most steadfast conformity. There was mile after mile of scrubland. There was the earth, a dusky red, and an occasional baobab as a happy change from all the stunted, wizened, smaller trees. There were also hornbills, red-billed, yellow-billed or black and white. And there were often hawks, perched solemnly as they stared at the land on every side. Although I never find any journey tedious, save for those at sea, the sameness did go on, through Zimbabwe, through Zambia, through the greater spread of Tanzania. For the first 5000 kilometres of its journey the Great North Road (so called, if not always particularly great) avoids the biggest mountains and comes nowhere near a lake. It crosses a river or two – the Limpopo, the Zambesi – and then travels through scrubland once again. As a result I was totally unprepared for Mount Meru, the volcanic mountain, which suddenly loomed ahead of me. Curiously it then vanished, leaving me happily bewildered as I rattled into Arusha, the most northerly town of Tanzania.

At dawn I realised I had truly entered another kind of world. Mount

Meru confirmed its existence by reappearing. To its east lay Kilimanjaro, a snow-capped mountain that looked even more astonishing in the early morning sun. Over to the west, so I learned, there were lakes, such as Eyasi and Manyara, and a most extraordinary escarpment. Beyond that sudden wall was Ngorongoro, an amazing caldera some 20 kilometres across and stacked with animals. Further to the west lay the Serengeti, with yet more creatures and thousands of grassy square kilometres for them to graze. As I travelled north from Arusha towards Nairobi there was similarly spectacular scenery, with every form and shape of volcanic effusion, the craggy outline of Longido, of Ol Doinyo Orok, and then of scores of lesser hills on Namanga's Kenyan side. After all the days and weeks of uniformity I and my bike had entered the Rift. The volcanoes were a part of it. So too the scarps and fissures, and the flat ash plains spread so widely in between. At one moment I and my machine almost vanished in a portion of this ash. It was entirely soft, being lighter than thistledown, and became all-enveloping following no more provocation than the arrival of a pair of wheels seeking for some firmness down below. Sometime in the past, and not too distantly, that finest form of dust had been vented from some portion of the Earth wishing to be rid of it.

So what was this Rift that everyone mentioned in response to all my questioning? Were there other rifts around the planet and, if so, where were they located? Why was this the Great Rift? What had happened in eastern Africa in particular to cause such heaving and apparent discontent in its portion of the crust? On travelling north again, having put Nairobi behind me, my questions multiplied. Why all these lakes, and why were some charged with soda while others were quite fresh? When did this rifting take place? Was it recent as geologists use that word or long, long ago? On speeding through Uganda, and meeting the Rift's western portion, my wonder intensified. When I finally met the Nile I could appreciate that I had left something quite extraordinary in my wake. This phenomenon had been the Great Rift Valley, the greatest such system visible anywhere on the surface of our planet Earth.

·I·

THE RIFT'S GEOLOGY

—

The sight of the Rift Valley, wherever an observer happens to be, can make that individual wish to become a geologist. A glance at the area's described geology, when compressed into a textbook, can make the same individual wish instantly for some more relaxing form of entertainment. The terminology may seem almost as complex as the Earth's features it is attempting to describe and explain. Nevertheless some understanding of a rift's nature can make the sight of the scarp even more exhilarating. This massive fault did have a cause. The circumstances leading up to it are explicable, and there is a sense in all the contorted confusion of every landscape, just as there is sense when dry earth or even a piece of pie is caused to crack.

The crust of this particular pie is the result of several thousand million years of cooling, heating, bending, twisting, compressing, stretching and general upset. Each new movement and change occurs on rocks already altered by earlier events. The layering of sediments creates a pattern which is then disturbed, being compressed and folded or by being pulled apart. Intruding through all of these layers are the volcanoes. They can pour rock or ash on to the land, or merely squeeze it beneath the surface layers. They can leave flat plateaus or high mountains, forming a variety of landforms as they spew forth, collapse, and find further weak areas to act as exit points, destroy those already made, and spew again an age or epoch later on. A piece of cooking can behave in similar fashion, venting here and there, oozing from down below, changing consistency according to temperature, but think of giving names to all that bubbling, layering, hardening, cracking.

A rift is defined as a cleft, fissure, chasm, rent, crack, split. A valley is a 'low area more or less enclosed by hills, usually with a stream flowing through it'. Africa's Great Rift Valley is all of those things save for the stream. There sometimes are streams, making use of the lower-lying land, but they have not formed it. The Rift Valley's involvement with water is more a chain of lakes than any kind of stream, but these lakes are by no means all interconnected. Instead it is the Rift that is more continuous, with water having collected at various points along its length.

More technically, according to a dictionary of geology, the term Rift Valley is a 'linear depression or trough created by the sinking of the intermediate crustal rocks between two or more parallel strike-slip faults … the structure is known as a graben and the accompanying morphological feature as a rift valley'. The German word *graben* implies a grave-like simplicity, with a lower portion between two vertical sides. Unfortunately the straightforward ideal, of a valley neatly slipped between two faults, is a rarity. What is more probable, and can occur most magnificently, is a series of step-faults. It is these that can be seen, from the vantage point of a balloon's basket, as they drop down from the Ngongs in Kenya to the Magadi area. Such steps form a major part of every portion of the Rift Valley. They may have been blunted, by erosion and by time, and may even be scarcely visible; but a fault is more likely to occur as a succession of lesser faults, each adding to the whole. The valley is in fact most changeable, with no portion of the gigantic system being more than vaguely similar to any other part. The land was varied before the rifting process started to tear it apart. Its differences were accentuated rather than made more uniform by the fracturing of the past 40 million years, the period during which most of the continent's rifting has taken place.

The Great Rift of Africa is not the only such feature on the Earth's surface. It is, however, the largest still readily visible, with the remaining recent major rifts all lying below sea level. A tremendous rift travels north and south in mid-Atlantic, forever increasing the gap between the New World and the Old. Another is in mid-Pacific, and yet another in the Indian Ocean. Linking these three is a fracture running from east to west, south of Australia, south of Africa, and through the tip of South

America. The Indian Ocean rift runs into the Red Sea, part of it heading north into Israel and Lebanon, part heading south through Ethiopia and then down through eastern Africa. The most conspicuous of these major faulting systems is, of course, the one travelling for the greater part of its length over land rather than ocean floor. As some two-thirds of the planet is covered with water, and as rifting leads to a lowering of the surface, it is to be expected that many of the major fractures are below sea level. It is therefore our good fortune that one exists where we can see it – the Great Rift Valley of Africa.

So why is the rifting happening? Why is such an aged planet, several thousand million years after its formation, behaving as if it has far from settled down? One answer is that the planet is not dead, without heat, without varying internal pressure, without movement anywhere. It steadily receives heat, unevenly, from the sun and then loses it, no less unevenly. It also creates its own heat, as radioactive elements within the depths of the Earth break down, sometimes into other radioactive substances, sometimes into more stable elements. The universe is also active, bombarding the solar system with cosmic rays, some of which affect our planet. As a result there are terrestrial forces that deform, that mould and shape the crust in particular. The word tectonic, after the Greek for carpenter, relates to building and structure in general, but has been adopted by geologists for the forces that create the shape and form of the crust. Tectonic movements compress the Earth's surface to cause mountains and, as their converse, tear portions of the crust apart to create rifting.

In 1965 J. Tuzo Wilson, the Canadian geophysicist, linked two existing ideas. One was that of continental drift, a concept first proposed in 1858 but actively developed in 1915 by Alfred Wegener, which suggested that entire continents have moved around the Earth's surface. The second idea concerned sea-floor spreading, which proposed that sea-floors grew from mid-oceanic ridges. Wilson's new theory of plate tectonics not only incorporated these previous proposals, but suggested that the Earth's crust consists of a series of rigid surface plates which move against each other and carry the continents along over a hot, partially molten lower layer. Such mobility implies correctly that the plates collide with each other, slide past each other and also move apart. The movements do not have to be continuous, and the East African Rift may be what is known as a failed arm of the Red Sea Rift. In general the continental rifts are failed attempts at plates splitting apart. Initially, when a rigid plate begins to experience the tensions that will split it, there may be several cracks. Eventually one crack takes over, leaving the others in the stage they had reached when the major crack began to dominate. (An example is the North Sea to the east of Britain. It tried, as it were, to become the Atlantic but failed during the Triassic/Jurassic when the present Atlantic Rift began to dominate. Had it succeeded, Britain would be even less a part of Europe than it currently considers itself to be.)

The Great Rift Valley of Africa is not an ancient occurrence, at least not in geologists' terms, and the start of the Tertiary Period is a convenient point to begin the Rift story. That was a mere 65 million years ago, a minute proportion of the planet's total existence of 4500 million years, give or take a thousand million. Life of some sort has been in existence here for some 3500 million years. When the Tertiary era began there were already mammals, fast taking over the supremacy enjoyed for so long by the reptiles. As far as Africa was concerned, there was certainly nothing of the Great Rift and not much in the way of mountains. The folded ranges of southern Africa already existed, but the Atlas in the north-west were yet to be formed.

The periods of the Tertiary era have been named, from oldest to youngest, as the Palaeocene, the Eocene, the Oligocene, the Miocene, the Pliocene, and the Pleistocene. There is considerable debate on the subject, such as when the Oligocene began and when the rifting began, but many believe that the faulting which resulted in the Great Rift started during the Eocene about 40 million years ago, although this did not happen simultaneously throughout its length and breadth. The Atlas mountains began to rise during the Oligocene, and during the Miocene much of the volcanic activity in eastern Africa occurred. During the Pliocene and Pleistocene the rifting continued and so did the volcanic activity. Humans emerged during these two periods, although crucial steps towards Homo's speciation had already taken place in the Miocene.

If the Rift began 40 million years ago this span of time is only about half as long as there have been mammals on Earth, or a twelfth as long as there have been fish. The time of Rift formation has therefore been short from the point of view of life on Earth. From the viewpoint of a single life, and of the movements to be expected in one span of three score years and ten, the period of rifting is unimaginably long. If a rock mass were to move at the scarcely detectable rate of 1 millimetre a year it would travel over 40 kilometres during 40 million years. The Red Sea is now said to be widening by about 2.5 centimetres a year, which would mean an increase of 25 kilometres in only a million years. This rate is hasty in geological terms, and the Rift Valley may still be developing. It is not tearing itself apart with the vigour shown on various occasions in the past, and the current situation could be a quiet phase marking the end of this piece of rifting. The Red Sea may be absorbing and succumbing to all the tensional forces in the area.

A much smaller feature of the planet than Africa's Great Rift, but equally well known, is the valley of the Rhine. It has a river running along it, but was formed in similar fashion. The fact that a river has taken advantage of this lower land is a result of, and not the cause of, the drop in altitude. Africa's Rift was not encountered by Europeans until the nineteenth century, and not by geologists until that century's last decade. Therefore plenty of opportunities had existed within Europe for such double faulting to be analysed closer at hand. A further major rifting caused Lake Baikal in eastern Siberia, the world's deepest lake at 1740 metres, but neither the Rhine region nor Baikal can compete for size with Africa's rifting, some 6500 kilometres from top to toe.

More correctly known as the Afro-Arabian Rift system, its northerly section lies in Asia, where it runs up the Gulf of Aqaba, into Israel, the Dead Sea, and the Jordan Valley. Northwards through Lebanon and into Turkey it grows less and less distinctive. Southwards from Aqaba it runs along the Red Sea, dividing Asia from Africa. Near Aden, one rift system branches to travel beneath the Indian Ocean, while the Afro-Arabian system enters Ethiopia at the Afar depression. This is an extremely hot, forbidding and shallow piece of land, quite different from the Ethiopian highlands lying to the south. In these mountains, severed by great canyons, the Rift is less distinct, but it becomes more conspicuous when entering Kenya and partnering the Omo river. A chain of lakes marks the Great Rift's passage southwards through Kenya and northern Tanzania, such as Turkana, Baringo, Bogoria, Nakuru, Elmenteita, Naivasha, Magadi, Natron and Manyara. Africa's largest lake, Victoria, is neither part of this chain nor of the Rift system, being midway between the Great Rift and the Western Rift Valley.

The Western Rift starts on approximately the same latitude that witnesses the decline of the Great (or Eastern) Rift. Some 700 kilometres separate the two of them, and a further lake chain marks the north-south passage of the Western Rift. These tracts of water act as borderland between Uganda and Zaire, then between Tanzania and Zaire, and finally between Tanzania and Malawi. The valley of the Zambesi river, in its lower reaches as it travels through Mozambique, is considered to be the ultimate portion of Africa's Great Rift. To travel from north to south in this fashion may imply that the fissure also rent its way southwards, but many geologists believe the rending occurred from south to north. It may have been a piecemeal affair, affecting first one area, and then another and another, but the prevailing opinion is that it was more systematic. In any case it has ended up as a crack, of varying degrees, running in a detectable fashion for much of its length. Almost all of the Rift, between latitudes 35° North and 20° South, lies between the 30th and 40th longitudes east of Greenwich.

Its precise width is extremely variable. Parts are relatively narrow, being no more than 40 kilometres from one side to the other. Others, less immediately recognisable as a rift system, are 200 kilometres or even 400 kilometres across. In Kenya, the valley is 40 kilometres wide in the area of the Kano Plains, 100 kilometres across at Baringo, 170 kilometres at the southern end of Lake Turkana, and 320 kilometres wide at its northern end by the border with Ethiopia. Plainly the zone of crustal extension, as geologists like to call the area of split, is not uniform. The tensions must vary; so too the rocks on which they are operating. Spread along 6500 kilometres, these differences are likely to be considerable, and it is therefore more amazing that an identifiable degree of uniformity exists over such a distance. Whether its width is 40 kilometres, 400 kilometres or scarcely recognisable, there is a continuity that is quite clear when viewed from space. The crack, and its varying width, and chain of lakes, present an astonishing phenomenon stretching for 55 degrees of latitude.

The fact that the Rift has resulted from faults can correctly suggest some kind of error, however natural the happening. In many places the Earth's crust is a regular set of layers, like the various fillings of some well-constructed cake. A fault disturbs this evenness, as when someone slices that cake to position a piece of it differently. The Earth's crust yields along a line, presumably where it is weak, and the layering is displaced as one side slips to a lower level. When this has happened the original layers are no longer continuous; the cake's jam has shifted to lie next to the sponge or cream. A fault can result from a tearing apart of the crust or a squeezing of it. Tensional faults are said to be normal whereas those caused by compression are said to be reverse. Within the African Rift system the faults are of the normal variety. There also exist tear faults, where the conflicting movement is horizontal rather than vertical, but these are not a major feature of the system in Africa. Instead it is a definite rending of the crust, with the forces pulling from both east and west to cause the north-south divide.

Africa's north-south scar is a complex system of faults and by no means a single happening. On small-scale maps it may seem that way, there being no scope for detail. On large-scale maps, however, it is clear that hundreds or even thousands of fractures have occurred, each adding to the whole. They can run more or less in the same direction, much as a piece of shattered wood splits roughly along the grain. A fault is often drawn as if it is always vertical but an angle less than 90° is much more commonplace. The text-book simplicity is in reality a multitude of odd-angled displacements, each complementing but obscuring the general picture. The land is split plain enough, but which bit came from where, and when and how, is nothing like so simple to observe, even with the aid of a multi-faulted map. Besides, apart from normal faults, and the reverse and tear faults, there may be hinge faults, oblique faults, pivot faults, thrust faults, and so on. That old concept of a *graben*, when two sides have a downthrow in between, can be entirely lost within the confusion of a rift system, and what makes everything harder to detect is that nothing happened yesterday. The fault may have been blindingly obvious when recently formed and may even have been vertical. Over the thousands or even millions of years its precision is eroded. Rocks tumble down the Rift. Streams also descend from the higher to the lower surfaces, wearing away the rocks as they go. Gradually the original incline is blunted and may even end as no more than a gentle slope.

It is exciting to imagine, when presented with some impressive, clearly defined, and beautifully conspicuous piece of faulting, that the feature was formed instantaneously. In reality there was not some amazing, frightening day in the past when an escarpment suddenly arose. Instead the fault may have appeared a metre or so at a time, with a series of such jerks spread over many thousands of years. In time the fault may be impressively high, but erosion begins to work the moment there is anything to erode, and the sharp clarity begins to be diminished long before the forces creating that particular fault have finished their task.

The big escarpments may seem to suggest that faults operate in areas already weakened by some earlier dislocation. If a fault exists, and some escarpment has resulted from that fault, it might reasonably be expected that the next such fault will add its displacement to the one already existing. Sometimes this does occur. On rarer occasions no one fault takes precedence and the region becomes a spilled matchbox of lesser faults spread across the land. Such a zone lies to the west of Lake Bogoria. To the east lies a magnificent escarpment, steep and everything a rift wall should be, but to the west lies much flatter ground, composed of a considerable cluster of little faults, sometimes called a fault swarm or grid faulting. North of Lake Magadi is a similar region, flat by comparison with major faulting but a topsy-turvy of lesser happenings. Had these little faults, with their diminutive throws, added their efforts together they would rival the great escarpments. Instead they spread their labour to make a swarm or grid. Sometimes there are faults every half-kilometre or so across the width of the Rift. As these are all of different ages, and have weathered differently, the task of sorting out what has happened, and when and where, can be considerable.

The sight of the big escarpments, such as north of Manyara or east of Bogoria, can also make one consider that surface faulting is always impressive, either now or at some time in the past. However there is no law compelling such fractures to be of any dimension. The vastness of the Great Rift suggests that all its component faults must also be large, but its considerable length is made up of very much shorter segments. Even the major features, such as those Manyara and Bogoria walls, only stretch for a few tens of kilometres, while the smallest faults can easily be overlooked. The observer has to switch his or her eyes from rift walls, perhaps over half a kilometre high, to look for features only centimetres high. Sometimes these minute slips can be seen in lava flows or cuttings where preparations for a road or railway have exposed a piece of land. The distance from one layer, perhaps of ash, to the continuation of that layer at a different level may only be the width of a person's hand. It is still a fault, however small. Some nearby pressure must have built up and caused, when the local material yielded, this minor slip. The Great Rift is the grand total of all such faults, huge and minute, that have taken place in its general area over many millions of years.

What has defined the Rift is the trend or strike of all these faults. Had they been haphazard there would have been no delineation, no superficial scar distinctive even to orbiting spacemen. They would have occurred randomly, proceeding towards every point of the compass. Instead their overall strike has been loosely consistent, being largely north and south. Some, notably in the region lying between the two main arms of the Rift, have a trend that is almost east to west; but, in the main, the faults can act as a rough (very rough at times) form of compass in this part of the world.

An astonishing fact of the most impressive Rift areas, such as those in Kenya, is that volcanic material filled up the valley almost as fast as it was created. Over in the Western Rift, this was not the case, its lakes being deep whereas all the eastern lakes are extremely shallow. During the past 30 million years hundreds of thousands of cubic kilometres of ash and lava have been deposited in the Rift region. In the central strip of the eastern Rift this lies 3 kilometres deep or more. Individuals who stand and stare at the awesomeness of the Rift should realise how inadequately the grandeur of the original faulting is being revealed. Lake Tanganyika's high walls may look impressive, particularly on the western side, but if the lake was abruptly drained of all its water those walls would suddenly stand 1500 metres higher. The lake's depth would have been added to the relatively small escarpment height.

The creation of a fault is always accompanied by an earthquake. The earthquake may be a single shock or a series of shocks in quick succession. A major fault, involving millions of tonnes of rock in a sudden movement, is bound to have great repercussions. It is a spasm, a convulsion of the earth. It may not add up to much, at least not visibly on the surface, but it is a dramatic release of energy, a sudden alteration of the status quo.

It is easy to think of earthquakes only in terms of the most dramatic. Japan, for example, experienced a 30-second tremor in 1923 that killed 143,000 people and destroyed 576,000 houses. Other names spring to mind where there have been terrible incidents – Mexico City, Managua, Agadir, Skopje, San Francisco, Messina, Charleston – but nowhere have people been killed more than in Japan and no occasion has been so destructive as that tremor of 1923. However the Earth does not only suffer extremely violent perturbations. Major movements are far less frequent than minor ones, and most tremors are insignificant. About a million occur in any year, with the vast majority detected only by the delicate instruments of seismographers.

The famous Richter scale was devised by the American, C.F. Richter, in 1935 to compare the magnitudes of different earthquakes by reflecting the total amount of energy released. The scale is logarithmic and so, for example, an earthquake of Richter 8 bears little relation to one registering Richter 4. The latter is barely detectable, save to a few people relaxing at the time, while the former will raze whole towns. The tremendous energies involved do need such a shorthand. Chile's horrific earthquake of 1960 is believed to have given off as much energy as is usually released by all the tremors over a normally active year.

Although it cannot compare with Japan the Great Rift is a region of earthquake activity. The Western Rift is much more earthquake-prone than the more famous Eastern Rift, and some notable earthquakes have occurred in the past. The first to be recorded by Europeans occurred in Kenya in 1892, at about the time the Rift Valley was being recognised for the feature that it is. In 1910 there was an earthquake of Richter 7.3 at the southern end of Lake Tanganyika, large enough to be felt in Nairobi and Mombasa. The second major tremor in Kenya was the Subukia earthquake of 1928, reckoned to have been Richter 7.1. Fortunately, the region of its occurrence, along the Laikipia escarpment to the east of Lake Bogoria, was (and still is) poorly populated. There was some damage to buildings, a few streams had their courses altered, and a 2-metre fissure appeared at the base of the escarpment near the lake.

In 1964 there was a quake of 6.4 down at Mbulu in southern Tanzania, sufficiently severe to be felt in southern Kenya and Uganda. Then came the most destructive earthquake in modern times to affect the Rift Valley area. At 4.45 in the morning of 20 March 1966, after some preliminary shocks for two days beforehand, there was a violent shaking in the Toro area at the southern end of Lake Albert (Mobutu Sese Seko). It was, and is, a well-populated area, with simple buildings. Some 6000 of these were destroyed, and 157 people died when roofs or walls fell on them. Fort Portal, in Uganda, lies only 50 kilometres away, and the loss of life would have been tremendous had the epicentre been that distance to the southeast. Fort Portal is acquainted with tremors, experiencing a dozen or so each year, but has so far escaped significant damage. It certainly felt the Toro event, but then so did much of Zaire, Rwanda, Burundi, and Uganda. Even in western Kenya, 650 kilometres from the epicentre, people were shaken from their beds. The location of the earthquake which so manifested itself in the Toro region was actually some 30 kilometres deep within the crust. Who knows what other pressures are building up down there now, and what earthquakes are likely to happen, destined to do even greater damage? Should they affect a town, say Fort Portal, Kabale or Masindi in Uganda, or Bunia or Goma in Zaire, the death total will be vastly greater than at Toro.

The future of the Rift Valley system is an even greater unknown than when or where the next earthquake will strike. Everyone agrees that the rifting has not come to a stop coincidentally with modern science's arrival on the scene. It has lasted for the past 30 to 40 million years and is plainly still occurring. Many of the lava flows are extremely recent, some even having faults in them. The Rift volcanoes are not yet dead. Many are still smouldering, and some – notably in the Western Rift – are busily erupting. There are still earthquakes and the Red Sea is widening. The Rift still lives, and this most conspicuous fracture on Earth is still yielding to whatever tectonic forces have been causing it to happen since, early in the Tertiary, it began.

Perhaps, however unlikely, it will split even wider, permit sea-water to rush in at the Afar depression, and eventually cut off the easternmost section of Africa to create a separate landmass. Madagascar was formed about 100 million years ago when a rift developed between it and the rest of southern Africa. Some 500 kilometres now separate the island from the continent. Something similar is unlikely to happen with the Great Rift, but whatever does occur will hardly take place overnight, even in the sense that geologists might use the term.

· II ·
VOLCANOES, LAVA AND ASH

——

There is something extremely appealing about volcanoes. Perhaps it is their abrupt emergence from the landscape, with no foothills to blunt their urgency. Maybe it is their shape which is so attractive, having a precision lacking in an ordinary mountain range. Possibly our knowledge of their fiery birth engenders excitement. There was probably not a slow growth, as in normal mountain building when layers of rock are folded into shape over the ages. More likely there was a speedy bulging, a dramatic increase of pressure, and then a sudden yielding as a piece of the Earth's crust ruptured in most cataclysmic fashion. In later years, however cool the volcano, however much clothed in trees or other growth, there is always a feeling that further devastation lurks below the surface. Will it happen again, soon, now? Has that steam vent, curling upwards in sinister manner, enlarged a little recently and is it a portent of tremendous drama? The word dormant has a nice ring to it, so much better than inactive. Perhaps this beautifully shaped, perfectly coned blister, currently tranquil, will abruptly tremble and then spew forth to add to that beauty, improve even that cone, or – Krakatoa-like – transform the bulk of itself into nothing more than dust.

For individuals who know that the rewards of walking uphill amply justify the effort involved, it is a particular pleasure climbing volcanoes, whether dormant or extinct. There is a simplicity to them, a neatness, and a certainty about the route. Mountain ranges can be confusing, with no apparent reason why one portion should be taller than the rest. A volcano tends to be an individual rather than a mountain crowd, making the spectacle from its peak more dramatic, with a steep drop down to the proper level of the land. An extra blessing of volcanoes is that they can give a two-fold view, inwards to the crater and outwards to all the world beyond.

The Rift valley is rich in volcanoes. The Virungas of eastern Zaire, which have arisen in connection with the Western Rift, include dead volcanoes, such as Mikeno and Karisimbi. Most of the volcanoes associated with the Great Rift are equally extinct. They have played their part and will not do so again. Others are merely biding their time, being dormant until the next eruption. Kilimanjaro and Longonot are two examples of this resting phase. In a third category are active volcanoes, and there are three areas within the Rift system where such activity exists. The violence may not be occurring now, or even for quite a while; but some form of eruption has happened in fairly recent times.

In the eastern region of Zaire are two active volcanoes, Nyamlagira and Nyiragongo, plus half a dozen now reckoned to be extinct, with Karisimbi – at 4507 metres – the highest. Kitazungurwa, the most recent to vent its form of spleen on the surrounding countryside, came from a small fissure associated with Nyamlagira. Many a volcano has several outlets, and all of the eruptions in the Virunga area throughout this century have been from lesser openings, such as three blow-offs around Nyiragongo in January 1977. In earlier years there have been effusions from Rugarama in 1971, Gakararanga in 1967 and Kitsimbanyi in 1958, all three linked to the mother volcano of Nyamlagira.

From July to August 1986 the fissure of Kitazungurwa sent lava through the forest of that area. Such outpourings act as a safety valve, releasing heat and pressure, and they can be approached, cautiously but safely, by individuals wishing to see a different aspect of their planet Earth. Although many animals can escape the flowing rock, the trees are either pushed over or left standing with, on occasion, no more than their upper branches protruding. As with any river, the rate of flow is dependent both on the volume involved and the lie of the land, but with lava there is an extra parameter. Water always behaves in a similar way, but lava can be either viscous and slower in its flow, or relatively speedy. It either rushes through the trees or slowly winds its way among them.

The second most active area of volcanic activity in the Rift region exists in the Afar depression, where the Great or Eastern Rift lies between the Red Sea and a portion of the Ethiopian highlands. This extremely forbidding region of the planet's surface, hot, dry and lying below sea level at its northern end, is a difficult place in which to live, with volcanoes providing an extra hazard. Erta Ale is one such vent, which was observed erupting in 1960 and 1970-71, but since the Afar depression and its Danakil desert are places visited only by the most intrepid, such as Wilfred Thesiger, author and traveller extraordinary, the Afar eruptions are poorly reported.

Far better known is the solitary member of the third region of Rift activity, Mount Lengai. It looks terrific, being in a particularly wild and appropriate portion of northern Tanzania, exactly on the line between two of the Rift's soda lakes, Natron and Manyara. The peak is surrounded by a supporting cast of extinct volcanoes, such as Gelai, Kitumbeine, Kerimasi, and Embagai whose own magnificent crater is filled with water. Lengai is not high enough for snow, but its summit half resembles some snow-capped alp because of the sodium carbonate emitted at every eruption, a rare substance for volcanoes. Its shape is everything that a volcano should be and its eruptions are exciting and explosive but not devastating. The last, in 1966, was the twelfth since 1880, and we may therefore be due for another. There must be underlying reasons why lava or ash is suddenly spewed forth from a portion of the Earth but, beyond saying that pressure or temperature have become sufficient to cause a rupturing of the crust, those forces remain a mystery. They choose their time and place, and we then learn of the event.

Lengai's 1966 eruption was much appreciated by photographers in light planes and most impressive to all who saw it. Minor activity has continued since then. January to March 1983 witnessed, according to one report, 'moderate quantities' of pale grey ash emerging from the crater floor. During May 1984 there were two vents on the floor's north side that 'contained a black bubbling liquid'. Three months later a 'dark grey to black, ropy, mud-like crust' had formed over the inside surface of the inner crater. By February 1985 a lake, 'boiling vigorously', covered a quarter of the crater floor. In 1986 there were 'three small peaks', each about 7.5 metres high. In other words, although without a major eruption, the volcano is far from dead. It simmers, it bubbles, it rumbles occasionally, and forever changes its appearance, particularly on that crater floor. Many believe that Lengai is due for another eruption, an exciting event for all save those present at the time with a most unenviable ring-side seat.

To climb Kilimanjaro is a considerable effort, often bringing on an attack of mountain sickness, and five days are recommended for the labour, three up and two down. Mount Kenya is not quite as high, but its tallest peaks demand rock-climbing, which decreases the enthusiasm of some potential conquerors. Lengai has advantages not possessed by these other two. It is an active volcano, rather than dormant like Kilimanjaro or dead like Mount Kenya. It can also be climbed in a day.

Lengai is much less tall, being a mere 2740 metres above sea level. The climb begins where the mountain begins, at 600 metres, and no connivance with a vehicle can knock off some initial contours. Vegetation has taken quite a hold, providing the earliest obstacle, and this later becomes elephant grass, not the easiest of materials. After the grass comes ash. Its hard, crusty texture is a refreshing change, but not when the crust repeatedly gives way, as feet punch unexpected holes in the surface. However, it is the gullies that are the major problem. These are steep, and can be very deep, making it necessary to walk along the ridges. Sliding into a gully, a distinct feasibility, forces a retreat until the climber can clamber up to stand upon a ridge once more. Worse still, a ridge can simply disappear when two gullies merge ahead of it. The trick is to choose a more tenacious ridge, one that reaches to the crater rim, but such a causeway is difficult to detect from the lower regions. With a good, reliable ridge everything is possible, and the determined climber can be standing on the rim an hour after leaving the last of the grass.

At this point comes visible evidence that Lengai is very much alive. There is a strong smell of sulphur and little wisps of steam curl from cracks within the ash. The steep slope of the crater leads down to all the weird shapes and protuberances on the crater floor. The rim is sufficiently wide to stand on without too much difficulty, but it is hot beneath the surface and there are fumaroles, or little vents, up there that are hotter still. To step into one, perhaps concealed beneath its crust of ash, is easy and unforgettable. To walk, slip, slide, stumble down to the crater floor is an occasional possibility, but there are times when such an act would be lethal, the ash being too hot, or too noxious in its fumes. The first European to climb Lengai, Professor Fritz Jaeger in 1906, was also the first individual to climb down into the crater. Currently most climbers are happy to stand on the rim and stare, to admire the spectacle from on high, to savour the odour from the Earth's bowels, and realise most forcibly that the crust on which life is possible sits most thinly on hot rock.

All along the Rift Valley are memorable signs that our planet has far from settled down, but none is quite so blatant as a volcano. As for climbing one, for looking deep within its maw, for seeing as much as can ever be seen of the planet's nethermost regions, this is yet more awesome and humiliating. The power down there quite outclasses anything on the surface.

Many of the Rift volcanoes are of the straightforward kind, with symmetrical craters forming the conventional cone. Known as central volcanoes, these erupt from the middle of the material built up around them. Active Lengai is one example; so too the less recently active Meru. Kilimanjaro would be another example, were it not for the fact that it is actually three volcanoes closely allied. Shira, furthest to the west, was the first to erupt and is therefore most eroded. Mawenzi, lying furthest to the east, was second and is still almost 5200 metres. Kibo, placed between these two, was third to blow and is highest, reaching 5895 metres, which makes it the tallest point on the African continent. This trio of volcanoes sits only 3° from the equator, causing initial disbelief in Europe that explorers had found an African mountain capped by snow. Mount Kenya, only 0°10′ from zero latitude, was also received incredulously when its iciness was first described, but in fact the equator is not the hottest region on Earth. Most of the world's deserts lie between 15° and 40° of latitude, existing where a high-pressure zone separates each temperate region from the tropics. However, there was a kind of effrontery about snow near 0° in Africa. It is still an odd delight to see Kilimanjaro glinting, to know there are glaciers up there, and to be standing on savannah sand as one views this contrary wonder of the hottest continent.

The spectacular volcanic peaks are memorable, but the profusion of lesser cones is also impressive. On the slopes of Kilimanjaro are some 250 small cones, perhaps about 100 metres high, perhaps much less. They can contain water, as with deep Lake Chala, or merely the ash and cinder they were busily exuding when their impetus died down. Longonot, so many people's favourite volcano due to the excellence of its shape and general availability, has a near-perfect extra or parasitic cone on its northern flank. As the mountain has a superb crater, a good recent lava flow to the north-west, steam arising from its inside walls, and a well-rounded parasitic cone, any chest-heaving walk up its slopes is not only refreshing but educational. Suswa, lying to its south, is more complex, having the appearance of an ancient hill fort possessing inner and outer ramparts. To create this shape the volcano erupted, slumped back within itself, erupted once again, slumped back yet again, and has been subject to considerable faulting so that a block of so-called lost land stands tall, and frustratingly inaccessible, within the central crater.

Very rarely is a volcano a single entity, formed once and then untouched. (No doubt archaeologists make the same point about hill forts being amended again and again throughout their history.) For one thing, as with all human constructions, erosion begins the moment it is made. A volcano's hardest rocks are often those which formed the central core when the activity died down. If the softer sides are worn away that plug can be left upstanding, a solid sentinel from the past. Mount Kenya is crowned by several such peaks, but they can occur seemingly randomly, jutting from nowhere in particular. One favourite stands guard to Hell's Gate, a game reserve lying south-west of Lake Naivasha. The upright legacy has a rocky cluster surrounding its base, ideal for a colony of rock hyraxes. Many will emerge to greet a visitor, particularly one who looks like giving food. Given less than half a chance, they bite the finger rather than the bread, not maliciously but certainly painfully (and I must better my technique).

The slumping back of volcanic material, as with Suswa, can occur either to a greater or lesser degree. It is possible to imagine a momentum that suddenly subsides, a great outpouring abruptly finished down below. Instead of pressure maintaining the eruption there is a falling away within, a collapse. If this happens on a large scale the entire upper portion of the cone can drop back into the abyss. Ngorongoro in northern Tanzania is the most famous such caldera along Africa's Great Rift. It is not the largest in the world, lying sixth overall, but is an impressive spectacle of an amphitheatre, with walls over 600 metres high enclosing 325 square kilometres. Many a major city could fit into that bowl; but, better by far, it is home for a superlative collection of animals. There is some forest, much fresh water and a mildly soda lake which waxes or wanes as it sees fit.

To gaze upon this wonder from the crater rim is to be easily confused. When I first looked, having been informed that the reserve contained thousands of large mammals, I expressed dismay that not a single animal could be seen. My companions countered with ridicule, pointing out the many dots that were wildebeest, zebras, gazelles, or the occasional ear-flapping elephant. My failure was not an inability to see the dots. It was a lack of appreciation of animal size when set within that vast encircle-ment. As its diameter is about 20 kilometres, and the circumference is a long day's walk of 64 kilometres, the caldera dwarfs its various features, such as the forest and the lake. It certainly dwarfs the animals for any untutored eye.

Later I attempted to balloon into this magnificent basin, keen on acquiring another perspective on all its animals. Having inflated by the rim we cast off shortly after dawn and caught the wind travelling across Ngorongoro. From a position that any eagle would envy we looked down into this incredible natural world. Needless to say we also wished to drop lower into it, and this we did by releasing hydrogen. However, no sooner done than undone. The old volcano spewed us out again, much as, some 3 million years ago, it vomited its lava to form part of what are now the Crater Highlands. We realised, after being despatched from the cauldron, that its enclosed quantity of air was being heated as the day was warming up. We were no more than part of the jettisoned excess. Expelled from the biggest saucepan of them all, we in our balloon were hurried over the crater rim much as milk hurries overboard when brought to boiling point. Once again I had not appreciated the size of this caldera, nor how much air it might contain. To warm such a volume even by one degree sends hundreds of tonnes of it rushing over the rim. To heat it up from night-time cool to day-time warmth causes the caldera to erupt, albeit with air, every single day. The early part of the day often provides one of the best of the many glorious Ngorongoro sights, with the sky a shining blue and the pan full to its brim with cloud.

The rim wall is steeper on the inside than the outside, as is customary with all forms of volcano. Even so there are places where the animals can walk up and over this natural boundary to their home. As fresh water is permanent, and good grass often grows in the crater when everywhere

Mount Lengai (right), the only active volcano in Tanzania, had its last major eruption in 1966. A satellite picture of Africa (inset) shows the extent of the Rift system, from the Mozambique Channel, through East Africa, into Ethiopia, and up the Red Sea into the Jordan Valley.

*Rift Valley escarpments (above) may be very high on one side, the
other being a series of smaller step-like ridges. Near Lake Baringo (above right)
a fault in the basalt has created spectacular cliffs. Rock hyrax (below right) are
found throughout the Rift Valley where rocky outcrops provide cover.*

Ngorongoro crater (above), the largest caldera in Africa, is 20 kilometres in diameter. Its crater has collapsed, but that of Kilimanjaro (left) forms the highest point in Africa. Others can contain water (centre left) or suggest continuing activity, as with Nyiragongo (far left) in eastern Zaire.

*On Mount Suswa long tubes were formed as lava cooled. These can
be entered where the roof has collapsed (right). Inside, on the walls, patterns
may be formed by splashes and drips of lava (below right). Sometimes the
lava can form beautifully symmetrical columns of basalt as at Hell's Gate
(below), near Lake Naivasha, where an overflow from an older lake
has spilled to erode this gorge.*

The Virunga mountains in eastern Zaire (below) *are
largely extinct volcanoes, but they are home to mountain gorillas* (right)
which live at altitudes of some 3500 metres.

*Mount Kenya (far right) is the second highest in Africa.
Its upper reaches experience severe night frosts, and near the
summit there are glaciers. Giant lobelias (right) are only found in
this Afro-Alpine habitat. They are pollinated by the scarlet-tufted
malachite sunbird (above) which feeds on their nectar.*

Massive outpourings of lava have produced the Ethiopian highlands
(left), *the most mountainous area in all of Africa. They provide a suitable
habitat for the rare Simien fox* (below) *which hunts singly, preying on
creatures such as the giant mole rat.*

45

At Djibouti (above) the three arms of the Rift Valley meet –
the Gulf of Aden, the Red Sea, and the Eastern African Rift. There
are many lava flows (left) in the highly active Afar depression, and
some recent cracks (far left) show that rifting is still occurring.

48

else is dry, the pressure to migrate does not exist and much of the population is resident. Indeed the place once became the curious residence of Adolf and Friedrich Siedentopf and their families. Before World War One they each built homes within the crater, one in the Lerai forest and the other 10 kilometres or so away. According to Bernhard Grzimek, the German naturalist, the two men cared for 1200 head of cattle by 1908. They also bred ostriches and tried to tame zebras. It is easy to be appalled that the men should have attempted to farm in such a perfect animal preserve, and therefore easy to be relieved that the British evicted them after hostilities began. It is also intriguing. How did they manage to grow crops with such a profusion of herbivores? How could they maintain any animals when surrounded by so many carnivores, from the largest lions to the more skilful leopards, and down through hyenas and servals to the even smaller kinds?

I once camped for a couple of weeks within Ngorongoro, and never failed to be enchanted by the place. Over my shaving mirror I would see zebras strolling past, so often sounding more like dogs. At night, and high within the fig tree above our tents, an animal would appear to be tearing itself apart, creating an eerie, heart-stopping cry, culminating in a throttled scream. I could not believe that any creature, let alone the gentle tree hyrax, could either make that sound or live to do it again night after night. Hyenas would whoop in sympathy, leopards would roar raspingly, hippos would produce their smug, deep-throated chuckle, and every now and then a lion would wind up its own powerful crescendo, much more like heavy panting than any kind of roar.

The crater floor is not of uniform height. The occasional plateau plus some rocky fragments serve as a gentle reminder that, once upon a time, eruptions happened here. These little kopjes offer trees for shade and a cool breeze to humans who have been in a vehicle too long. The fact that animals tolerate mechanical invasions of their territory is an enormous blessing, permitting a closer range of observation, but the animals never extend such laxity to anyone who leaves the privileged four wheels. On

A line of volcanoes (above left) in Djibouti continues under the sea as volcanic islands. The raised reef on the nearby coast (below left) may be the result of a geological upheaval or even a drop in sea level.

one occasion, I felt in need of a cooling change and, after leaving the vehicle by a suitable kopje, strolled up a path that led towards the trees.

The roar and the sight were simultaneous. The experts say that one should stand one's ground, remain immobile, become master of the situation. In fact, I turned immediately, long before appreciating why I was turning. The lion had no need to leap with such full-throated energy; a diminutive snarl would have been sufficient. However, it did leap, it did roar, I did turn, and for some unaccountable reason I did not have the beast land upon my back. Perhaps it saw the others, also strolling from the vehicle. Or perhaps, in sending me pell-mell with such immediate and insolent ease, it had performed its task of despatch entirely adequately. We all took our seats swiftly, and would have driven off with similar panache but, unfortunately, one detail was proving troublesome. The key had been left in the ignition, but I could not make it turn. It was a simple act but quite impossible. My fingers still quivered uncontrollably.

'I'm sure I'll be able to turn it soon,' I said.

'Well, fairly soon,' I added later.

No one forgets their time spent in Ngorongoro, meaning 'extra down' in Masai, whether or not they are carelessly casual with its largest predator. Before its cone collapsed this volcano may have been as high as 4500 metres, and the thought of such an enormous quantity of volcanic material falling back within itself is almost as daunting as the idea of such a cone being created in the first place.

The Rift is currently at a quiet volcanic period, and apart from the three major regions – Virunga, Afar and Lengai – there have been few outbreaks in recent times. Mount Meru, which either dominates the town of Arusha to the east of Ngorongoro or vanishes completely within a covering of cloud, flexed its muscles moderately in 1910. Presumably its altitude of 4566 metres will be amended by future eruptions, as the volcano is not considered to be as dormant as Kilimanjaro standing even taller only 65 kilometres away.

A few lava blocks from Meru, one or two additional streaks down Lengai plus a modest cloud of ash, and some hot rock flowing around Danakil or through trees in Kivu province does not add up to much when compared with some of the massive outpourings that have occurred in the past. Much of the Rift Valley would be well below sea level, perhaps 1000 metres below in places, were it not for all the volcanic material lying on top of the original surface. In Kenya this layer is occasionally 3000 metres deep. On the Ethiopian plateau, where there has been the maximum deposition of lava, the thickness is over 4000 metres. This material did not appear in one or even several tremendous eruptions. Instead the depth has been repeatedly increased, with each lava flow perhaps no more than 3–5 metres thick at a time.

It is the sense of continuity, however slight, which is such a stimulating feature of the Rift. All of the area's volcanic activity has happened in the past 40 million years, or so it is currently believed. The great lava plateaus of Ethiopia, for example, were formed in the first half of this period. Even a non-geologist flying over this area can appreciate the sight, not only of such a depth of volcanic material but also the gorges that rivers have carved through it. Much of Ethiopia is one grand canyon after another, sometimes running almost parallel. Villages on each canyon's opposing edges may be no more than a kilometre or two apart but might as well be distanced by an ocean for the amount of casual traffic that can take place between them. The Blue Nile (or River Abbai) has been particularly vigorous, cutting the old plateau to a depth of 1400 metres. Few will therefore walk, or climb, between the two sides, the drop down being as great as the highest point in Britain.

To fly north from Addis Ababa is initially to gaze upon an even carpet of land, with little clusters of round huts scattered indiscriminately over it. Then begin the gorges, one after the other, each hundreds of metres deep. The huts and their fields reach to the edge of each abyss and, that passed over, a similar gathering of huts and fields starts again on its other side. Life is surely strange, staring across at your neighbours, seeing their children play and knowing whether or not they have gathered in the year's harvest, but never speaking with them. The area is like an inverted Alpine range, with the inaccessible portions being the valleys rather than the peaks. After one short hour of flying, it is easy to understand why this region poses major problems for relief agencies. Swift transport by road is often out of the question because roads in turn are very frequently out of the question. Ethiopia is a beautiful, exciting and often difficult country. Its geography makes it so.

The feeling of continuity between the past and the present is enhanced by knowing which lava flows have been laid at which times. Ethiopia's basalt, the commonest form of volcanic material, is extremely old and looks it. So too is the rock of much of north-western Kenya, having been poured over the land about 15–20 million years ago. Further south, from

Maralal towards Nairobi, the dominant lava is a mere 12 million years old. The capital sits on volcanic rock put there some 3 million years ago. Further south again, in the Magadi area, the lava is about a third as ancient as below Nairobi. A particularly exciting volcanic phenomenon not too far away is the Yatta plateau. This finger of lava runs south-east from Thika for three hundred kilometres. It is always narrow, never exceeding 10 kilometres in width, and now stands relatively tall above the surrounding countryside. Plainly its lava was of harder material and less likely to be eroded than the neighbouring land, and this harder flow probably made use of an existing river valley. Since then, due to the erosion, it has ossified that ancient waterway. What happened there is as plain as many of the extremely recent outpourings. These can be spotted from the air or encountered, and then stumbled over, on the land. Some vegetation may have caught hold or the new rock may be too young for that to have occurred. Some of the material looks as if it poured forth yesterday, that is, until the observer sees some that actually did flow, say, a couple of years ago.

With lava plateaus 40 million years old containing gorges carved in them ever since that date, with volcanoes that exploded 20 million years ago alongside others that erupted in 1966, with tongues of lava still detectable that hurried across the land when humankind was still apekind, and with lava that killed a forest only the other day, there is a strong sense of succession that runs from the early Tertiary to modern times. This is true, of course, for the planet's crust as a whole, which has been eroded and raised, buffeted and ground down, throughout geological time. However, there is an extra dimension to the vulcanicity of the Rift. It happened then, and is happening today. There was lava; there is lava. There were volcanoes that are now but stumps of their former selves. There are volcanoes still producing, still adding to their bulk. It is a continuum, with former times and modern times most clearly intermingled.

One bizarre extra of a volcanic environment is lava caves, and there are particularly fine examples around Suswa. At first it seems inexplicable that lava should form tunnels, perhaps a score of kilometres long, when pouring down a mountainside. However, watching such a flow and seeing it cool makes it possible to comprehend the caves. The lava loses heat most on its upper surface and so begins to form a solid roof over its river-like progression downhill. Gradually that cooler roof becomes complete while the hotter lava continues to flow, now shielded from the cooling air. Finally, due to changes within the volcano creating the lava supply, this flow ceases at its upper end. The hot rock hurries on downhill but leaves a steadily increasing gap behind it. The lava will flow on and eventually congeal, but leave behind it kilometre upon kilometre of empty tunnel to mark its passage from crater rim to the flatter land below.

In theory such tubes through the lava are covered throughout their length. In practice, time and erosion having done their work, it is possible to find entry points, such as the one a group of us encountered. Disguised by an enormous fig tree, and surrounded with scrubby acacia, a major opening in the ground led down to the start of the cave. Unlike limestone caverns, probably carved smooth by water, a lava tunnel is a rough affair. Walking over surface lava is never easy and it is harder underground, darkness being an added problem. We had a torch but to concentrate its beam upon some awkwardness beneath meant disregarding a further hazard sharply overhead. We stumbled, we occasionally crawled ignominiously and we certainly hit elbows, scalps and knees, while the bats hit nothing but fluttered as they passed.

It was as well we used that torch to find what lay beneath our feet. At one frightening moment we saw that nothing lay beneath them, our progress being not so much along a floor as upon the roof of another tunnel existing down below. Further probing with the torch revealed yet another black-edged tube below that tunnel. Roofs were also floors, showing that lava had frequently been belched from that volcano to pour downhill, perhaps a year after some previous flow, and then on top of it, perhaps a millennium later still. Those of us probing in the dark, wondering how the bats survived such an odour, even if it was of their own making, were remorselessly astonished at this further aspect of a volcano's capabilities. It was like investigating the volcano itself, although horizontally. Would there be magma round the corner, the basic lava stuff? Or a boiling vat of rock, biding its time? Or was that seeming lack of air affecting us mentally, creating hallucination, stretching imagination much too far? We thought it time to leave, and stumbled, hit, cracked, jarred and manoeuvred where we had done these things before. The blueness of the sky is always a treat. On that day, and on emerging from the cave, we thought it better than ever.

Later, when still near Suswa and well within the Rift Valley, I watched as a herd of cattle was completely enveloped in its own cloud of dust. The animals were not hurrying, or kicking at the ground, but the air looked choked, as if creatures at full gallop were stirring up the earth. I wondered how the animals could breathe, and wandered over for a closer look. Soon I too was leaving a substantial cloud, as my feet walked silently through a depth of dust I had encountered only once before. On that earlier occasion, when on a motorbike rather than two feet, the whole machine had disappeared with not much more than a woof of noise into an apparent sea of talcum powder. The engine stopped, resenting the sudden lack of air, and the motorbike changed colour instantaneously. So, I suppose, did I.

On both occasions I had been made to realise that volcanoes do not always emit lava; they can also fire out ash. This never pours, it simply rains down following a violent explosion. The magma, pulverised or shattered by the blast and mixed with gas or steam, is hurled upwards. The finest ash powders can travel for many kilometres and may land only to be blown elsewhere. Lapilli are larger pieces of ash, more like pebbles, and so fall nearer to their point of origin. Some blocks may be huge and these travel even shorter distances. A volcano's rain can also include lava bombs. These have a rounded shape, possibly twisted at the ends, and are formed when magma is sent spinning while still warmly malleable. Pumice, yet another form of lava, is like a solid sponge and so filled with air that the rock will sometimes float. All such items, thrown upwards and landing in their own good time, are known as pyroclasts – broken by fire – or tephra after the Greek for ash.

If these materials, notably the lightest forms, are still hot on landing, they may become fused to form a more solid piece of matter known as tuff. This can be of many colours – black, brown, green, white – and a yet greater variety of texture. It all depends upon the original composition, the size, the temperature on landing, and the pressures exerted afterwards. The magnificent Serengeti plains, home of the greatest concentrations of large mammals anywhere on Earth, are basically formed of a fine-grained, light-brown tuff. This is thought to have originated from Kerimasi, a now extinct volcano, situated over to the east. The wind still often blows from east to west, just as it was doing when Kerimasi blew its top and sent its brand of ash westwards to form the Serengeti tuff. Tree roots tend to find these ash plains impenetrable. As a result, superficial grasses can flourish, and do so abundantly. The volcanic soil is extremely fertile, and only dry years can lead to a scarcity of food for the millions of animals that live on the Serengeti plains. Kerimasi did a magnificent job all those years ago.

Even the volcanoes themselves may be home to some astonishing animals. The Virungas of eastern Zaire are, among much else, a suitable location for the mountain gorilla, the blacker, shaggier and rarer sub-species of this great ape. It was not until the twentieth century that a European, Captain Oscar von Beringe, first saw *Gorilla gorilla beringei*, and he promptly shot two of them. The animals were discovered even after the shy, elusive, forest-dwelling okapi had been found, gorillas also being shy, elusive and forest-dwelling but living at altitude. This height, and the generally difficult terrain, makes the business of finding the Virunga gorillas not only an exquisite pleasure but totally exhausting.

Our party of three started its ascent of Karisimbi in the Virungas from a village at about 1500 metres. It was then a walk up, and the villagers pointed out the way. Much of the path had also been a stroll down for elephants, a fact the locals omitted to mention, and over the enormous skid-prints we slid most readily. To escape the path we entered the forest, and its entanglement made the path desirable once more. Even halfway up, when plunging through the bamboo and carelessly treading on fallen lengths of it, when having my feet depart from the rest of me most instantaneously, I knew I had never been so exhausted. A volcano ought, or so I reasoned, to be covered with lava or at least ash; but there had been ample opportunity for earth to be created and then, with the daily rainfall, for this to be fashioned into mud.

In time, most magically, the contours relented. We entered a lush meadow, thick with grass, dappled with sunlight. It was the saddle between Karisimbi and its volcanic neighbour, Mount Mikeno, and was a most delectable spot. My body, awash with sweat, ached for rest, this longing being happily satisfied where the grass grew longest and the sun dappled at its very best. On the way up we had encountered the *Hagenia* trees, and had dropped ourselves over their conveniently low-slung branches, rather like some washing suspended on the line. That was pleasurable, hanging there and observing the place much as a sloth must see the world, but nothing could compare with the cool perfection of being cosseted by grass.

The sun then set, or rather it disappeared behind one slope of Mount Mikeno. The vanishing did not immediately remove all light, but warmth departed instantaneously. From being hotly moist with sweat I became chillingly cold and wet. A discarded shirt was found at once, as was another shirt, a jersey, and then a suit brought for smart Nairobi evenings, but all proved inadequate. The rucksack exhausted, I tried to generate heat by running, only to trip over a considerable grave. It belonged to Carl Akeley, a much-travelled American naturalist. He had died within two days of arriving at that hot-cold grassy plateau back in 1926. I shiveringly wondered if he had had the good fortune of an extra suit or two. It did seem an easy place in which to die, however dressed.

We slept that night in half a hut, and half-slept, appropriately. The construction was a research post but, during political upheaval, had been ignited in some form of protest. It seemed entirely comprehensible, in a place where heat could change to cold so suddenly, that the conflagration had lost interest in itself and, like poor Akeley, had expired leaving half a hut untouched. We were extremely grateful for the surviving portion, cooked in it, and eventually lay down in it, with half a heaven of stars to see and wholly convinced we were warmer within than without. At that spot the sun not only set early but, due to the landscape's shape, rose early. And so, without much effort, did we. After breakfast, having stepped outside, we were ready for the day. In particular we were ready for gorillas.

The procedure was straightforward, or so we had been informed. We would walk until we found some tracks. We would follow them until they looked fresh, made within a day or two. Then, more cautiously, we would travel along these until, in theory, we would encounter the gorilla group that had made them. In practice, there is a problem in detecting whether 7-day-old signs are fresher than 6-day-old signs. The vegetation, half-eaten or merely crushed by the passing animal, can look similarly bruised. The nests, made both for night-time and a mid-day rest, can appear equally ancient to the uninitiated, whether they have been constructed in the crook of a tree or are lying on the ground. The faeces, always deposited in each nest once it has been used, can quickly sprout mould, thus blunting the sight of the droppings themselves but providing an extra form of clue. It is very possible for the pursuer to track backwards, having been misled by the varying signs. Worse still, another gorilla group may have passed that way, intermingling the tracks, causing the tracker to retrace steps again and again.

The animals are never daunted by a steep incline. They go up, down, along, across, as the spirit moves them. They do not find intervening trees an inconvenience, using them much as a human might welcome hand-rails placed along a path. Anyone in pursuit of a gorilla group finds them one more obstacle as he climbs over and under the branches, endlessly clambering up or down, and wondering how far a single group can travel in a day. There are false trails, where a gorilla or two has sauntered from the main path. In time the erring animals retrace their steps to the main group, thus forcing the humans, wearily, to follow suit. All the while, as tracking and back-tracking proceeds, there is total awareness that gorillas, the largest apes of all, are near at hand. Since the mature and silver-backed males reach 180 kilograms and their strength is known to be phenomenal, there is every reason for anxiety, save that the animals very rarely attack. Occasionally they do mis-read a situation, considering their trackers to be offensive or believing them to be a threat, notably to their infants. Normally they see no reason for any more belligerence than a considerable yell and much chest-slapping. Only when this has proved inadequate, or when they are suddenly frightened, will further action be initiated.

An additional problem, apart from the thinner air of mountain slopes above 3000 metres in height, is that neither trackers nor gorillas can see each other. The vegetation, so tailor-made for steadfast consumption by gorillas, is often 2 metres high. The trees are thickly hung with mosses, further obscuring the view. The animals know they are being followed, even though they cannot see their pursuers, and occasionally send forth a slapping sound. The gesture of chest-beating is often mocked by humans, but there is nothing silly about that noise when it comes at you out of the silence of the wild. It is no simple slap but an awesome, hollow, intimidating clatter which does everything the gesture was meant to do. It subdues, it humiliates, and informs every living thing that a gorilla is nearby.

Within a couple of hours we found the nest-beds made for the previous night. The deposited faeces were clearly fresh. The wild celery had plainly been recently chewed, and the knuckle-prints had the clarity of a potter's finger-marks in clay. Then, suddenly, the air was rent apart by one of the loudest noises I had ever heard. It was a scream, a yell, a shout, all intermingled into one most terrifying sound. The cry seemed to fill the air like thunder. It gave no hint of its point of origin, save that its maker was very close. It completely stopped us in our tracks, and for a while I wondered if I would ever move again. The others indicated that we should retreat a little towards a convenient tree. I managed to turn round and, with heart still pounding heavily, moved slowly after them.

A *Hagenia* proved easy enough to climb, even for one whose limbs were still quivering. We had been informed that nothing should be done, when in a gorilla's presence, which might be construed as aggression. Apparently any sudden gesture, any lifting of field-glasses or raising of cameras, can be misunderstood. We climbed slowly, stretching gently upwards to find another branch before, with unhasty movement, pulling ourselves to a higher level. We made no noise. We did not even speak.

As we climbed, as we lifted arms in turn, we gradually saw on a neighbouring tree that another male individual was behaving in identical fashion, climbing, lifting his arms one after the other, and also gaining altitude. We stopped and he stopped, with each side staring at the other. We had shirts and he a hairiness, but there was a likeness of primate and primate, ape and human. In time this other individual settled himself more comfortably, and in time we three did as well. With field-glasses unraised, with cameras still encased, we gazed across at him while he seemed to gaze across at us. There was no certainty because, although his nostrils and lower face were visible, the eyes were deeply entrenched and overhung with hair. Once he stood up and slapped his chest to make that sound again. It was less terrifying now that we could see its source, and I was intrigued to observe that his chest was bare in the place where male humans are most likely to have hair.

Other arms began to climb into the trees, causing other bodies to become visible. These mountain gorillas were a dark black, a total contrast to the vegetation, but they still managed to blend into its unlit recesses, making it quite difficult to keep track of them. As one slipped out of sight, another appeared momentarily, or one of them tossed leaves high up into the air. Once the big male made a sudden charge, scattering his fellow gorillas as he did so, and again we three were unnerved. He then wrenched out plants, and tossed them skyward leaving a few still hanging in a tree. Finally, almost casually, he retreated from his front-line stance, moved slowly to the rear, and prepared to make another charge. This he did, most effectively, from back to front before ending precisely where he had begun. On this occasion we were more intrigued than alarmed, having observed the subterfuge involved.

The three of us stayed fastened to our trees, and were rewarded with a most pleasing hour or so. The big male, so much the largest of the troupe, relaxed his displays and settled down to eat. The females and smaller males also ate in a more casual manner, hiding less frequently and letting us get to know their different faces. The youngsters behaved as only youngsters can, irritating their elders, running along branches, falling, and one even putting a leaf upon a head. It stayed in place for a while, fell, was replaced, and filled us with the enchantment of the scene. Eventually the group moved off further up the slope. We too then descended from our trees, stretched stiffened limbs, shook hands happily with each other, and retraced our steps. As we finally left Karisimbi, forever looking back at the two volcanoes with the plateau of Kabara lying in between, we felt we had visited a most friendly, alien land. *Gorilla gorilla beringei* had been our host, and we had enjoyed that company quite tremendously.

Plainly, to everyone who has ever visited the Great Rift area, there is an association between the rifting and volcanic activity. However the link is obscure. Many volcanoes are at quite a distance from the Rift, such as Kilimanjaro, Kenya and Elgon. Many portions of the Rift have no volcanoes, either near or far, such as the western areas around Lake Tanganyika and the southerly Lake Malawi. Although it is easy to imagine a fault line as an escape point for magma, this is not true. The rocks slid past each other so tightly that no opportunity was provided for the release of magma. Volcanic material surfaces nearby, and is undeniably linked with the rifting, in that volcanoes and rift occurred contemporaneously; but the creation of a rift does not necessarily mean the creation of a volcano, any more than a volcano creates a rift. It so happens that whatever forces down below lead to rifting lead also, on occasion, to eruptions.

The escarpments, fault lines and other features of the Rift are all impressive, but no more so than the amount of volcanic material linked to them. Geologists have calculated that, in the Kenyan and Ethiopian portions of the Rift Valley, over 400,000 cubic kilometres of lava and ash have emerged in fiery or gentle outpourings to pour or rain upon the land. Such a quantity would cover all of England to a depth of over 3 kilometres. It is therefore amazing that anything can now be seen of the Rift, following such effusion. However, both processes are still in action today, the rifting and the vulcanicity, the faulting and the erupting, and no one has the glimmer of an idea when the double act will end.

· III ·
'POCKETS' GREGORY

——

John Walter Gregory was born in Scotland in 1864. More than anyone else he put Africa's Rift Valley on the map. The Rift had been seen before his arrival in eastern Africa, but it had not been properly described. To say he was a geologist is correct but inadequate. An official bibliography of geologists goes a touch further in calling him 'explorer, stratigrapher, structural geologist, geomorphologist, and invertebrate paleontologist'. In fact many more titles could and should be added, such as botanist, author, diplomat, adventurer, global traveller, and total enthusiast for encountering geology *in situ* rather than in books. The fact that he died in 1932, aged 68, in the turbulence of the Pongo de Mainique on the Urubamba river high up in the Peruvian Andes shows that his interests did not relent merely because he had gathered some years along the way.

His three main autobiographical works help to describe the man, even in their titles. *The Great Rift Valley, being the narrative of a journey to Mount Kenya and Lake Baringo, with some account of the geology, natural history, anthropology, and future prospects of British East Africa* was published in 1896. Ten years later came *The Dead Heart of Australia; a journey around Lake Eyre in the summer of 1901-1902, with some account of the Lake Eyre basin and the flowing wells of Central Australia*. And seventeen years later there appeared *To the Alps of Chinese Tibet; an account of a journey of exploration up to and among the snow-clad mountains of the Tibetan frontier.*

He also made notable travels in Libya, Angola, Spitzbergen and India. Nearer home he was equally indefatigable, writing several of his 300 scientific papers on the Tweed valley. Those Andean rapids hurrying down to meet the Amazon cut short the life of a man plainly with more work to do, more countries to visit, and papers to be written.

Before Gregory's investigations in eastern Africa scientific examination of the area had been rare. However, in one way or another, the Rift Valley and its various related features had been noted. Europeans had seen Kilimanjaro in 1848, Mount Kenya in 1849, the Western Rift with Lake Tanganyika in 1857, and the southern end of the Great Rift, with Lake Malawi (formerly Nyasa), in 1859. When Baron Carl Claus von der Decken, one of the best of the traveller-explorers, was killed by Somalis at Bardera on the Giuba river in 1865, European enthusiasm for opening up this world waned temporarily. There were also Arab maps, rich with facts collected over the centuries, but their considerable information was scornfully neglected by Europeans. Gregory makes the point that Lake Bogoria (formerly Losuguta and then Hannington) was well known to Arab caravans but did not appear on European maps until 1887 when a European had actually seen it.

Towards the end of the nineteenth century the prevailing European view was that Africa was one great slab of metamorphic rock. Geologically it was a continent, or so they believed, without a history. In 1852, Sir Roderick Murchison, president of the Royal Geographical Society, put forward his opinion that Africa south of the Sahara was a land of great antiquity and simplicity. Nothing much had happened there for the last two of the three major eras of the Earth's history. Even in 1891, the year before Gregory set out, Professor Henry Drummond wrote 'The thing about the geology of Africa that strikes one as especially significant is that throughout this vast area just opening up to science there is nothing new – no unknown force at work; no rock strange to the petrographer; no pause in denudation; no formation, texture, or structure to put the law of continuity to confusion'. When people dig holes for themselves they often make them nice and big, and Professor Drummond was no exception. Africa was to demonstrate a tremendous force at work, with the world's greatest rift on the surface of its land.

Drummond should have listened more to the explorers who were actually observing the features of the Rift. Baron von der Decken's death only created a hiccough in African exploration. By 1883 Dr Gustav

Fischer was examining lakes Natron and Naivasha, thus exploring the Rift Valley floor in some of its most exciting areas. Joseph Thomson (after whom the Thomson's gazelle was named) braved the Masai, went further north than Fischer, mapped Lake Baringo, and was the first European to see Mount Kenya from the west. His view of the Rift was therefore considerable, not least the Laikipia escarpment which provides such a magnificent eastern wall to Lake Bogoria.

In 1887-8, as a further powerful contribution to European Rift knowledge, came the hunting expedition of the Hungarian sportsman, Count Teleki. With him travelled the admirable Lieutenant von Höhnel, an Austrian who made notes, maps, and recorded events other than each day's bag. This expedition saw the lakes of Basso Narok and Basso Ebor, renaming them Rudolf and Stephanie. (They are now Turkana and Chew Bahir.) According to Gregory, the map made by von Höhnel was 'so precise and instructive that, with the aid of the author's descriptions and sketches, and a small collection of rocks, it has enabled a very satisfactory account to be given of the geology of the area'.

Professor Eduard Suess of Vienna came to conclusions about the Rift Valley that were never achieved by those who, pre-Gregory, had actually walked over it. He proved, without moving from his Austrian study, that the chain of lakes along the line from Syria down to Nyasa were due to a connected series of earth movements. He even attempted to date them, being firmly of the opinion that Africa was a continent *with* a history. Gregory overflowed with praise for this distant observer. 'With his usual insight into geographical problems, he has read more of the lessons of the country, from descriptions, than the travellers who wrote them did from the country itself'. The task Gregory set himself was to be both a Suess and an explorer, a geologist who would see the results of the connected earth movements and understand them at first hand.

He had his first experience of eastern Africa at the age of 28. This was an unhappy experience, and he prefaced his account of it with the Swahili phrase *kulekeza si kufuma*: to aim is not to hit. The journey had started in precipitate fashion and others might have let the opportunity pass. Gregory only heard about an expedition due to investigate the least known portion of the Rift Valley four days before it left England. As he was then an employee of the British Museum of Natural History he needed permission to vacate his post for many months at least. He also had to collect equipment, and eventually set off one week after the others had left. Fortunately, by leaping on board departing ships, he caught up with the party at Aden.

Gregory and some others from the expedition started upon their exploration by striking north from the Kenyan coast along the Tana river, but within a few weeks Gregory was desperately ill. Two of the party died, and others would have done so had not a fitter section of the expedition made a timely arrival. Fresh meat, 'all sorts of European dainties', and considerable rest revived Gregory's strength but erased his memory of that terrible time. This expedition was destined not to succeed, and Gregory eventually arrived at Mombasa determined to try again. He found comfort in another Swahili proverb, 'To lose the way, that is to know the way.'

Later that same year of 1893 he left Mombasa, this time in charge of his own detachment. On this second foray, instead of proceeding north along the swampy and therefore malarial Tana basin, Gregory struck north-west. It was the more conventional route, taken by Arab traders over the centuries, followed by pioneers of the Imperial British East Africa Company, utilised (later) by the railwaymen as they made tracks for Uganda, and now the course of the Mombasa-Nairobi highway. He intended to follow this route and then branch off when he came within reach of the Rift Valley system.

Gregory appears to have got on well with his men, who called him *Mpokwa*, meaning 'Loaded pockets', because he was generally weighed down with samples culled along the way. To raise his prestige – or so was the intention – he predicted a partial eclipse of the sun. His Greenwich Almanac forecast the event and the sky looked sufficiently clear for it to be observed. To disbelieving ears he affirmed that a bite would be taken from the sun shortly before it set. At the very hour when his prophecy should have been vindicated, causing heaven knows what reaction from his men, a thick bank of cloud appeared from nowhere. Gregory explained to ears, now steadfast in disbelief, that the bite had indeed been taken, if only they could see it. There was an understandable and most logical reaction. 'If Mpokwa can bite a piece out of the sun, he could have kept those clouds away.' For days, whenever clouds appeared, they asked him to bite sections out of them. Prestige had not been enhanced, but friendliness had probably advanced by way of compensation.

An early port of call was Tsavo, followed by the fort of Machakos, founded four years earlier. An Englishman was in charge and the two

Britons spent time discussing place-names. They agreed that local nomenclature should be preferred to titles proposed by the first Europeans to come that way. If places 'are called after eminent people at home' these 'are useless, being unknown in the district ... and merely burden geography with a set of synonyms'. Such conversations, of expatriate with expatriate, would be in short supply after Machakos. Fort Smith, slightly further on, possessed the last European face Gregory would see for quite a while. It was also the place for some final shopping, including a sack of vegetables, 'a small flock of sheep', three donkeys (for portage rather than consumption) and 'a ton of other food'.

Suitably equipped, the expedition continued on its way. The local Kikuyu howled and jeered but did nothing more alarming. It had been five weeks since the start of this second safari, and as the famous Rift Valley grew closer, Gregory became more and more impatient. As luck would have it, the sight he anticipated proved elusive. 'Just before we reached the summit of the pass, a dense cloud settled down upon us and completely blotted out the view.' Fortunately, as often happens in that area, the cloud only sat like a cloth on top of the escarpment. By the time donkeys, sheep and men had descended several hundred feet 'a wonderful prospect burst upon us. We were on the face of a cliff 1400 feet in height, broken only by a platform 500 feet above the floor of the valley'.

Gregory and his men looked across, from somewhere near Limuru, to the great cone of the volcano Longonot. To the south they admired 'the breached crater Doinyo Suswa'. Beneath them was 'the dark sinuous line of flat-topped acacias' that mark the course of the Guaso Kedong' and over in the west 'the long, dull gray scarp of the plateau, which forms the western boundary of the valley'. The donkeys, less intrigued by one of the world's great views, took the opportunity to throw their loads and bolt. Everyone pursued them and, on the way, Gregory noticed old shore lines on the Rift Valley wall. He realised they were lake terraces, this valley having once been covered with water. Promptly (and without

Lake Nakuru in Kenya (right) *is a shallow Rift Valley soda lake, known for its vast population of lesser flamingoes. Although they come to feed here they do not breed on the lake.*

*Greater flamingoes gather at Lake Natron (above left), a most
favoured site for breeding, with Mount Lengai as a superb backdrop. Both
greater and lesser flamingoes spend time at Lake Bogoria (below left), with its
steep escarpment and hot springs (above) where drinking water is available.*

Greater flamingoes take to the air (above) and an adult pair tend their single, day-old chick (left) on Lake Elmenteita.

Lake Natron (left) may be favoured by the flamingoes, but its concentrations of soda (below) make it lethal for most other creatures. Humans will have their skin burned and may even be blinded should they spend a long time there.

The River Omo (below) *flows from the Ethiopian highlands into
the Rift Valley and Lake Turkana.* Egyptian geese *(middle right) and*
white-faced tree ducks *(below right) occur widely in eastern Africa, as does the*
African jacana *or* lily-trotter *(above right) whose long toes enable it
to walk on floating vegetation.*

The Turkana (above) *belong to a traditionally pastoralist tribe, and have only taken to fishing recently.*

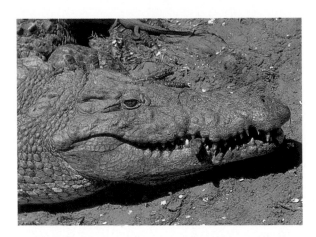

Hippos (below) *generally feed at night and spend their days in the water. Nile crocodiles living by river banks, such as this female carrying young in her mouth* (above left), *tend to eat more meat than lake crocodiles* (middle left) *who consume more fish. A terrapin family* (below left) *is seen by a small stream in the Masai Mara reserve.*

Lake Turkana (below) *contains several crater lakes which are most suitable for a wide variety of birds. Yellow-billed storks and spoonbills* (above left) *favour many of the Rift's lakes. So do white pelicans* (below left) *which can migrate from as far as Europe.*

contradicting the Machakos conclusion, as the spread of water no longer existed) he named the huge and former lake after Suess, the Austrian professor he so admired.

The valley floor was not Kikuyu country and the expedition could therefore relax its military style. 'Free from the bother of the caravan, I could climb a mountain, track a river, visit a neighbouring lake, chase butterflies, and collect plants as careless as a schoolboy'; but that carelessness was nearly Gregory's downfall. Having agreed to meet up with his party that night by 'yonder hill' he hurried off, with some of the most exciting geological country in the world to explore. Principally he examined 'Doinyo Nyuki', which he discovered to be 'the denuded remnant of an old volcano'. He traced the bands of ash and tuff that encircle it, climbed its four highest pinnacles, and saw – or thought he saw – his camp being pitched on that yonder hill. Casually he sauntered down Nyuki's slopes in good time. Equally casually he wandered, in fading light, towards the camp site. At sunset and with alarm he realised the camp was not where, in his opinion, it should have been. Worse still there was no cooking fire visible in any direction. Rain then added to the darkness, and Gregory remembered how one of Joseph Thomson's men had been killed and eaten by a lion in this same area.

Seeing a lion at the foot of every bush, he decided to find the camp, come what may. Fortunately he encountered a stream and knew the tents must have been pitched by its bank. Unfortunately the stream meandered and sub-divided, causing much back-tracking on his part, until the geologist was lucky enough to observe the unmistakable signs of sheep, donkeys and Zanzibari porters. The darkness and rain forced him on his knees as the only way of following the marks until, after a considerable distance of this humiliating procedure, he saw a light ahead. Men were out searching for him, but he ridiculed their fears, for a man who attempts to bite pieces out of the sun does not get lost. 'I did not think it necessary to add that the feelings of self-reproach and fright that had possessed my soul more than outbalanced the charm of the scramble over the peaks of Doinyo Nyuki'.

The Awash River cascades down the Awash Falls (above left) *on its way to Lake Abbe where the water disappears in the searing heat, some evaporating and some sinking underground. Volcanic steam vents leave extraordinary formations of dried mud* (below left).

Next day he was out on his own again, examining the terraces of Lake Suess, and sheltering from the rain in a deserted Masai kraal with two decomposing bodies to keep him company. On another day, he determined to climb Mount Longonot, its excellently shaped crater admired today by everyone passing by road or rail from Nairobi to Lake Naivasha. Although 2780 metres above sea level, the actual climb from its base is less than 920 metres, with much of this being a stroll or scramble round its rim. Joseph Thomson had climbed only to the rim in 1883, and Gregory determined to do rather better by reaching the peak, the high point on its rim, and collecting information on its geology. Noticing that the mountain's lower part consisted 'of a series of platforms or terraces of lava' he found the rock to be a 'black trachytic pumice, containing a good deal of obsidian'. Further up 'the great surprise was the discovery of a large steam vent on the inner face of the north wall of the crater'.

That steam vent is still there to inform modern strollers and scramblers that Longonot is a volcano, dormant rather than extinct. Gregory was reminded of this at every turn, whether encountering a narrow sharp ridge of volcanic ash or staring down the steep slopes that lead both internally and externally from the crater's rim. He ascended the final section 'hewing a way through the scrub with a sword bayonet' until he could stand upon the summit. It is a superlative view, well worth the climb, and Gregory settled down to sketch it. Later he boiled thermometers, a simple system for determining altitude, and computed a maximum height only about 70 metres higher than subsequent measurement.

The next stop in this itinerary of Rift Valley examination was the Mau escarpment on its western side. Gregory wanted to discover if its structure corresponded with that of the eastern wall, which would prove that the two sides had been pulled apart. Unfortunately a group of Masai warriors stood in the way. He asked them if they were elders (*el moru*) or merely warriors (*el moran*). On being told, as he presumably knew, they were young men, he became instantly scornful and sent them off to bring some elders. This they did, but the *el moru* were equally unhelpful, demanding that he return in the direction from which he had come. Gregory, having been informed in Mombasa that bluff was the only suitable expedient, said he would march on through their territory. They then demanded *hongo*, a kind of toll, and suggested a figure equal to all of Gregory's possessions. Both sides retreated and the geological party, suspecting imminent attack, built a considerable fence or *boma*. The Masai's required friendliness, cemented by a handshake or, better still, a spit, was plainly not forthcoming for the time being.

In the morning heavy rain arrived and with it the elders, who repeated their demands and stressed that caravans 20 times the strength of Gregory's pitiful assortment had occasionally been massacred. Allegedly there were 9000 warriors around Naivasha. Gregory said that there was a great caravan coming that way from Uganda, whose men 'were more in number than there were papyrus stems around Lake Naivasha'. He assured them that if the Masai killed him, this force would 'sweep upon their country, kill all the *el moran*, eat up all the cattle, and drive the elders, women and children out into the deserts where not even their slaves, the Wanderobbo, could manage to live'. Following this speech the Masai drove their women and cattle back to the kraals. 'It looked most uncomfortably like a fight.'

In fact, the bluff worked. When the rain stopped, Gregory broke camp and the Masai elders held out a 'knobkerry, cut from a rhino horn,' for him to shake. Gregory did so, but knew this kind of meeting was of little consequence. He strode on towards the lake and was then offered, for better measure, a hand. The two men held each other's hands, at first coldly, and then more cordially. Gregory had been told that, without a spit of peace, future squalls could be expected, and was therefore greatly relieved at the eventual expectoration. It had cleared the air for further geologising.

Some 100 kilometres further north, the expedition reached a high-spot of its Rift adventure. Gregory and his men had marched north past the Aberdares, where they had repeated encounters with rhinoceroses. They had then gained in altitude as they headed for the Laikipia escarpment east of Lake Bogoria. While trying to find a good camp-site Gregory found himself 'without the slightest warning . . . on the edge of a precipice 1900 feet in height'. The party had reached one of the most magnificent spectacles of the entire Rift Valley. He painted it so vividly, was so ecstatic in his prose and founded so much of his explanation for the rifting upon this general area, that a major portion of the famous valley is now known on geological maps as Gregory's Rift.

'At our feet, at the base of the precipice, lay a long narrow lake, in shape something like Windermere. The opposite shore is formed of a series of steps and terraces which rise one above another to the summit.

Beyond this ridge is the valley . . . and beyond this again are the undulating foot-hills and the dark gray scarp of the plateau of Kamasia. The view was certainly the most beautiful that I had seen in Africa'. Praise indeed from someone who had already looked down from Kikuyu country to Naivasha and beyond, who had seen Longonot rising with such magnificence from the valley floor, and who had had the double pleasure at its peak of looking both outward and inward, either beyond the crater or within its steep and tree-clad walls.

The expedition had been running short of drinking water and perhaps the sight of all that Bogoria water made the view yet more enchanting. At all events they hastened to reach it, but soon realised that the view, however excellent, was not without its drawbacks. Although the precipice was steep rather than sheer, it presented a considerable problem for descent. One man found a little water in a few rhinoceros footprints, and this was eagerly collected, but no convenient route downhill was discovered during the remainder of that day. After a breakfastless start on the following morning, the scientist enjoyed himself with 'some interesting plants and a most instructive geological section'. Having climbed for a distance downwards he came up against 'a vertical cliff of lava'. The escarpment does not look so formidable today, but Gregory was accompanied by donkeys, sheep and considerable head-loads. A good path was therefore necessary and an enterprising porter, Fundi Mabruk, located a game track leading to the lake. Unfortunately, by then, a second day had almost gone. Another night of rhino-imprint water was necessary above the tantalising spread of Bogoria.

Once again, breakfastless, they made an early start, happily knowing that all would soon be well. When the descent had been completed one man hurried forward and drank deep and carelessly before shouting 'Dowa, dowa, hapana maji'. It was not water, he concluded, but medicine, and he soon demonstrated its emetic qualities. The donkeys, unimpressed by this violent proof that Bogoria was yet another soda lake, had to be constrained by sacks placed over their heads. The sheep, strangely, were permitted to drink and, within a short time, two of the remaining three had died.

Gregory should have suspected the water might be undrinkable as he had seen, from above the escarpment, 'vast flocks' of flamingoes 'floating on its surface'. He had already encountered 'the putrid sulphurous water of Lake Nakuru' and had seen that 'whatever the water touched it seemed

84

to kill'. The flamingoes at that earlier lake had been in such numbers that 'one of the kite-shaped flocks ... measured 400 yards in breadth and a mile in length' and these 'alone seemed able to touch the waters of the lake and live'. If flamingoes desire the algae that thrive only in such soda circumstances, it might be expected that they would not congregate on water sufficiently fresh for humans to consume.

The expedition was now not only lacking in water but desperately short of food, the two dead sheep notwithstanding. Fundi and Gregory set off to shoot a rhinoceros, having seen a pair in the vicinity. There was a convenient stream-bed up which the men could stalk the single individual that had most recently shown itself. Unfortunately, its partner had also favoured the dead ground and did not hesitate to charge. Gregory was able to fire, but his barrel was so choked with sand that the recoil was tremendous. The bullet merely passed through the rhino's muscular neck, and the hunter gained no more than a severely damaged shoulder. Later that day, Gregory spotted another rhino and suggested another hunt to Fundi. 'Enough rhino for one day,' was the reply, a reminder that *fundi* is the Swahili word for expert, as in wisdom or even caution.

So the expedition continued hungry and thirsty round Lake Bogoria, seeing tantalising glimpses of rain over to the west. When they eventually found good water, in correct African fashion, they had it to excess. At the north end of the lake there was sufficient for all their needs and much more than they wanted. In the middle of a refreshing swamp flowed a river, too deep to wade, too swift to swim. Worse still it contained crocodiles, and even walking along its slippery banks became too dangerous. After days of scouring rhino footprints for something to drink, the expedition now had to build a bridge over an unwelcome torrent. The men felled and hewed timber, working through the night, and were rewarded after a safe crossing by finding *maji moto*. These hot springs on the western shore of Bogoria permitted the entire caravan to 'lark about ... like schoolboys', as many travellers must have done before and certainly since. The trick is to discover a point where the bubbling, gushing sulphurous water has cooled sufficiently, and where there exists a slight depression in the steaming stream, to make a perfect bath. In fact it is more than perfect, offering as it does, a view of hundreds of thousands of flamingoes, the lake multi-coloured by all its algal growth, and that magnificent Laikipia escarpment, which Gregory had carefully descended, along with sheep, donkeys and porters.

As the journey continued northwards from Lake Bogoria, it encountered starving villages and there was threat of mutiny. 'Bulging pockets' managed to hold his band together and also persevere with his work, mapping Lake Baringo, collecting, and eventually reaching a point of return when some 145 kilometres south of Turkana. 'It was mournful to have to turn back when so near (the lake), but it was useless to go on unless I could get some time there for scientific collecting. To have dashed across the desert, simply for the fun of dashing back again and saying that I had been there, would have been an unjustifiable waste of energy, and a needless risk'. Resources were presumably running low; so too the patience of his caravan; and so, one imagines, the patience of his employers back by the Cromwell Road.

He made time for Mount Kenya, however, and was the first geologist to examine it. Other Europeans had seen it, climbed much of it and even collected from it, but, in Gregory's view, unsatisfactorily. Compared with Kilimanjaro, the studies of East Africa's second great, snow-capped volcano were extremely modest. Even its name is due, he claimed, to a misunderstanding and should be either Kilinyaga (Kikuyu), Doinyo Ebor (Masai, the white mountain), Doinyo Egeri (Wanderobbo, the spotted mountain), or Njalo (Wakamba, even if the name is also used for Kilimanjaro).

One problem for the traveller is that Mount Kenya, like other volcanoes, can give no trace that it exists. A mountain range has foothills, rushing rivers and other clues to its whereabouts, even when its higher regions have vanished into cloud. Mount Kenya simply disappears. Dr Ludwig Krapf, a missionary, was the first European to observe both the mountain and this phenomenon. On 3 December 1849 he saw the peak for a few minutes from a distance of 145 kilometres. He stayed in that same district for several weeks, never saw it again and was not universally believed. As Kilimanjaro had only been observed by a European during the preceding year, and that report of a snowy peak so near the equator has been much ridiculed, Dr Krapf's slim evidence brought further mockery. Krapf returned two years later to the same observation point at Kitui, some 130 kilometres due east of Nairobi, but Mount Kenya was, once more, nowhere to be seen. He advanced towards the alleged peak for 65 kilometres, still failed to see it, and was finally forced to retreat by a belligerent party of Wakamba.

Another German, Hildebrandt, also spent time at Kitui 28 years later but he never saw the mountain. Snow near the equator was no longer in dispute, for this had now been accepted for Kilimanjaro. It was the existence of the mountain itself. Arab and Swahili traders spoke of it, but the Europeans dismissed their reports. It was Joseph Thomson, first in so much, who affirmed the mountain's presence for European disbelievers. He not only saw its western face from Laikipia, but correctly surmised that the peak 'represents the column of lava which closed the volcanic life of the mountain ... The crater has been gradually washed away'.

Others then examined it at closer range, notably representatives of the Imperial British East Africa Company in 1889 and 1891, and Count Teleki's most able Lieutenant von Höhnel in 1892. By comparison Kilimanjaro had been, according to Gregory, 'visited by more than a hundred Europeans and carefully explored'. It was more accessible, and could even be temptingly seen from the coast. Estimates of Mount Kenya's height varied from 5500-7000 metres, and so it was – possibly – even higher than its rival 320 kilometres due south. (In fact Mount Kenya is 695 metres lower than Kilimanjaro's 5895 metres, all the estimates in Gregory's time being on the high side.)

By 1895 the scene was therefore set for a roving geologist to make a proper examination. Gregory's preparations for the ascent went ahead in earnest. Twelve men were selected to climb the mountain. Ten days' rations, extra clothing and reserve food were packed, and the ascent of Mount Kenya began. Unfortunately, as he had suspected from the start, Gregory could not achieve the summit. Nevertheless he did spend considerable time at altitude, being particularly intrigued by its effects upon him. 'Anything like undue haste was followed by an attack of loss of breath ... Every step upward required a distinct and painful effort ... It was not a mere feeling of lassitude, but as if my legs were paralysed, and my left foot numbed.' Since no one in Europe could achieve such a height – Mont Blanc being only 4810 metres – there was considerable interest in mountain sickness among European climbers. Gregory claimed two visits over 5180 metres. With his normal care he examined his own physiology as well as the mountain where, for once, he could only 'work in spurts'.

The mountain was eventually climbed in 1899, but there was no real need for Gregory to do so. The surviving peaks of this decayed volcano were no more informative than the remnants slightly lower down. Moreover his porters, who had to carry the inevitable samples, did not welcome clambering about on snow. On the first day of the descent Gregory found that his sickness and lassitude diminished and by the next day he had fully recovered.

'Pockets' of the Rift Valley had completed his investigations in eastern Africa and returned to Britain. He had already spent time in the Rocky Mountains of the western United States, and realised parallels could be drawn between 'the vast expanse of undulating prairie' west of Machakos and the lava fields of Idaho that he had seen from the Tetons. He felt certain that he had been seeing in Africa 'a plain of lava and not of alluvium'. It ran over the land, and into the hollows of the mountain 'just as the water of a lake follows the irregularities of its shore'. Such lava seas were known elsewhere, apart from portions of the United States, and it had been proposed by the geologist Baron von Richthofen that molten rock caused them after it emerged via fissures from subterranean reservoirs. Gregory thought little of this fissure notion for Idaho, and even less of it for East Africa. 'Nowhere could I find any sign of the fissures.' What he did find, and in abundance, were 'a vast number of small craters' and 'dozens of the denuded stumps' of craters generally flattened since the time of their activity.

He therefore proposed a theory to account for the great quantities of East African lava that was midway between cone eruption (with a major volcano emitting lava) and a fissure eruption (with lava emerging from cracks). He suggested plateau eruption whereby a sheet of lava was formed by numerous scattered small cones, these appearing where lines of weakness crossed each other. This opinion was reinforced by the fact that 'wherever these great lava plains are known, they occur on high plateaux composed of rocks which either retain their originally horizontal position, or are of remarkable uniformity in composition'.

As for the Rift Valley area, he stressed that all the valleys lying between it and the coast were made like those in England, 'for their courses were sinuous and their slopes rounded'. But when his party emerged from the Kikuyu forests, and looked down into the Kedong valley, across to Longonot and Suswa and right across to the Mau escarpment, 'we entered one which was straight in direction, and was bounded by parallel and almost vertical sides'. He knew that the Kikuyu and Mau escarpments had once formed a continuous plateau, and must therefore be grabens.

Despite Gregory's earlier approval for Murchison's speculation he then, most gently, turned the latter's opinions upside-down. 'Thus, instead of

88

there being no new type of structure in this region, and the geological facts being of wearisome monotony, it teems with novel problems, and all its conditions seem different from those of Europe'. Murchison's statements did not coalesce with Gregory's exuberance for the brand new experience of eastern Africa, where 'valleys are often due to rifts instead of to erosion; the mountains are sometimes formed of blocks instead of by folds; while the lava flows are on a scale that shows the impossibility of measuring the universe by European standards.' On a rough map of the area drawn for his book he shows the extent of this volcanic surfacing. The area stretches in a broad band from south of Meru and Kilimanjaro up to Mount Kenya and the Mau scarp, before vanishing northwards.

In his book, Gregory made considerable attempts to classify the various geological happenings of the country through which he had passed. He believed that the rifting was largely due to periods of volcanic activity followed by earth-movements to restore equilibrium. He argued that, as masses of volcanic material had been piled up on the surface, and as the subterranean reservoirs had surely been emptied, the 'upper layers of the Earth's crust were therefore over-weighted'. Hence a period of relative tranquillity after each outburst while the land subsided to create a better symmetry. He thought there had been three principal phases: the Kaptian, after the Kapte plains; the Laikipian, after the Laikipia escarpment that lies on eroded surfaces of older lavas; and the Naivashan, after the lake, when eastern Africa was much less arid and great lakes – such as Suess – existed that were '400 feet above the present floor of the valley'.

His terms are rarely used these days, his notions having been superseded by others, including the present-day theory of plate tectonics. Later geologists have had more time, more opportunity and better knowledge of other areas. They have also not been beset by the many problems he encountered, and the need to proceed everywhere on foot. However it is good to note that almost all of them refer to the central portion of the Great Rift Valley – the Kenyan section that leads south to Tanzania – as the Gregory Rift. He was not the first to tramp that region. However he was the first to study it in detail and most emphatically he put it on the map. How right, therefore, that his name should also appear on maps. For him there was the rest of the world to see, until his work was tragically cut short by the turbulence of the Urubamba river in Peru.

One wonders, of course, what filled his pockets then that may have helped to drown him.

·IV·
CHAIN OF LAKES

—

Without the Rift Valley there might be no lakes in the eastern half of Africa. As it is, about 30 of them provide some of the area's principal delights. They form a double necklace, with a string of western Rift lakes running parallel to the main eastern string which leads from Ethiopia southwards to Tanzania. In general the lakes are extraordinarily dissimilar. Some shimmer with soda. Others are blatantly deep. And some can vanish in the dry years, leaving only a plateau of hardened mud to mark the spot. They may be ringed with trees and themselves heavy with vegetation, or they may appear to be no more than a spread of water which, at a certain level, becomes arid land. Generally the eastern lakes are long and narrow, mimicking the Rift's own ribbon-like dimensions, but a few have shapes nearer round or square. None are long and thin from east to west – that would not be in the nature of the Rift.

There is, nonetheless, an overall simplicity to this double necklace, blurred only by the erratic progress of the River Nile. This longest and most powerful of African rivers arises in the general area of the Rift but seems disdainful of the fact. Its two principal sources, Lake Tana feeding the Blue Nile and Lake Victoria the White, are not Rift lakes. Their locations must have been influenced by the compressions and tensions partnering the creation of the world's largest surface crack, but they do not lie within it.

Lake Tana's Nile follows the more straightforward course, as it hurries away from the Ethiopian Highlands. Admittedly it travels south-east initially, and therefore further into Ethiopia, but within 160 kilometres it curves first south, then west, and finally north-west, as if remembering the desperate need for water within the vast Sudan. The White Nile is more bewildering, mainly because it arises between the two halves of the Rift. It plunges determinedly enough from the only point at which Lake Victoria spills off excess, now tamed by the Jinja hydroelectric station, but then loses virtually all sense of direction. It meanders to Lake Kyoga, another non-Rift body of water, before wandering in a more purposeful fashion towards Mobutu Sese Seko, formerly Albert. Confounding the earliest geographers the river enters this lake from the north and then abruptly departs northwards a few miles further on. After this brief dalliance it abandons the Western Rift and forges steadily northwards, finally linking up with Ethiopia's Nile at Khartoum. Just as the Rift's lakes do not, in the main, feed rivers, so do the rivers of the region, notably both Niles, detach themselves from involvement with the Rift.

The watery specks of blue on the map, which lie within the Rift, start with a scatter of lakes around Djibouti. Gargori, Gamarri, Affambo, Abbe and, further east, Assal mark a separate faulting leading eastwards from the main north-south line. The next major lake grouping, lying to the south of Addis Ababa, includes Zwai, Shala, Abaya and Chamo, all on the Rift's true line. Then, come Chew Bahir and Turkana, fed by two different rivers. Three islands of dead volcanoes in Lake Turkana, known as North, Central and South islands, make a splendid addition to the scene, as does the old volcano in the middle of Lake Logipi south of Turkana in the Suguta valley.

Further south is a dry valley pocked with volcanoes, and then Lake Baringo with more volcanic protuberances, one of which curiously, and enchantingly, resembles Pooh Bear floating on his back. Half the size, but no less beautiful, is Bogoria, followed by the even smaller Lake Nakuru and then Elmenteita, smaller still, sometimes to the point of vanishing altogether. The next in line, Naivasha, is thought by some to be the most exquisite of them all, with Crescent Island – part of the rim of an old volcano – projecting perfectly into it. Magadi, well to Naivasha's south, is more soda than lake, but is undoubtedly fascinating. So too the far bigger Natron, beloved by flamingoes but fearsome for any humans who think they too can strut about in a similar manner. The necklace continues, with Manyara, Eyasi and Kitangiri, but there is a divergence here, and it is the Western Rift that then takes up the thread.

It starts more or less where the White Nile hurries in and out of the lake called Albert by the British, this being appropriately diminutive (6% in area) relative to Lake Victoria. Southwards, and linked by the Semliki river, are lakes Edward and George (the latter becoming Lake Idi Amin Dada for the 6 years up to 1979). Next in line comes Lake Tanganyika, the world's second deepest (only beaten by Baikal in the Soviet Union), and only slightly bigger than Lake Malawi, which visually carries on the Western Rift and is Africa's second largest lake (Lake Chad not really entering the lists as it wanes in area so substantially each season). Southern Africa, below that river mouth, is all the poorer for not sharing in this most excellent necklace. The string of lakes, totalling over 100,000 square kilometres (therefore half as large again as Lake Victoria), is all the more fascinating for being such a mixture of different entities.

Take Lake Manyara, for example. On one day in particular it nearly took me, so confused was I between its placid beauty and its awesome properties. From high up, along the escarpment that runs along its western shore, the spread of water looks enchanting. Anyone driving on the road which leads from the village of Mto wa Mbu towards the crater of Ngorongoro in Tanzania will pause when the magnificent Manyara view emerges above the trees. It is the Rift quite splendidly displayed. Above the valley walls, often from a bluff some 30 kilometres south of the road's observation point, there can be tremendous cloud, delicately beautiful, spectacularly big. Africa's cumulo-nimbus clouds arise from nothing, from the clarity of a bright blue morning and can be 20,000 metres from top to toe by the afternoon. Their turbulence can only be imagined, with air being hurled upwards for 20 kilometres or so as lightning flashes and thunder booms to give due warning of the activity within. From outside they look like cotton-wool, but no resemblance is more deceitful. They are lethal, as only clouds can be.

Our intent was to fly the balloon over Lake Manyara. We would fly south with the prevailing wind, and pass over that amazing lake before being forced to land, either on ground high above the lake or within the valley that it occupies in part. It had been a wet year, and the spread of Manyara was greater than normal. Consequently, which may seem difficult to accept, the lake had become less easy to observe from the valley floor. A thick bank of reeds had sprung up, particularly on the northern shore, and all we saw of Manyara before the flight was that long-range view from the escarpment road. We were therefore exhilarated when, having laboured all morning inflating our balloon with hydrogen, the basket rose from the ground. The sight was better even than we had imagined. Buffaloes were in those reeds, flicking their ears and caring nothing for our presence not too far above. There were also hippos, idling the day away, but the greatest spectacle was the lake itself, stretching southwards seemingly for ever. From our modest height the escarpment looked even more magnificent, more formidable, and a superb backdrop for the lake. The sky was a translucent blue with not a hint of cloud.

We were carried southwards, precisely where we wished to go. There is an enclave known as Endabash, where a stream from the upper plateau has remorselessly eaten at the cliff. The balloon, as if following our wishes of the time, wandered in and out of that recess, showing us elephants, waterbuck and giraffes. At first these tallest of all animals looked like poles pushed over from the vertical, until we saw that the giraffes were sitting on the ground, legs tucked beneath them while their long necks made pole-like shadows on the earth. Then we were over the lake again, staring at the whitened contours where thin lines of soda marked earlier levels on the land. Manyara is not as alkaline as Natron or Magadi, but it too has no outlet and is steeped in soda. Its liquid is most unpleasant, certainly to drink and even to touch, for it has an alien, soapy feel.

Suddenly, causing us all to grip on to the wicker, a gust came from nowhere. There was no warning before we were being twisted this way and that, with quite a different feel of air blowing over us. The altimeter received all our attention as the needle travelled round the dial, showing a rapid rise. Our sizeable shadow on the lake shrank steadily, and soon we could look over the cliff wall at the plateau receding from its upper edge. Quite soon, very soon, we were very high above the lake, our dot of a shadow being quite invisible. Also impossible to see, but nevertheless all too plain, was the cloud creating itself above us. Our huge envelope masked the sight of what we knew was happening: a cumulo-nimbus in the process of formation. Bad luck had dictated that we were at its base precisely as it started on its momentous career to heaven knows how high above our heads.

The Rift Valley, formerly only beautiful, had now taken on a most sinister aspect. Its escarpment was acting as a trigger for the thermal that would end as a massive thundercloud. We might be scooped up by all that rush of air to become a piece of cu-nim. Failing that, with the cliff so very near, we might merely be thrown at its rock. Or, if the gyrating wind thought otherwise, we would escape its grasp to drop into the lake far below. The maps had told us that Manyara was some 50 kilometres long by 16 wide and I knew that a swim of 8 kilometres, assuming a central drop, would be far more than a match for me. Swimming for only a short distance in such an alkaline solution, with sensitive eyes, mouth and nose being the first to be attacked, would have one spluttering, and then gasping, and then – it seemed most probable – giving up the ghost.

As we started descending, I jettisoned two cautious handfuls of sand to try and maintain a midway station between the deep blue lake and the cloud above. Within a minute the descent was stopped and we were climbing up and up, the world becoming darker while the air grew colder. From our increasingly perilous eyrie we were being given a most memorable and dramatic angle on the Rift. Hundreds of metres of escarpment, sheer in places, was all too clear. The fault-line running north and south was most conspicuous. As we climbed and fell, more like a yo-yo than any flying machine, we were given repeated glimpses of the upper plateau, stretching to the west. The higher ground received the majority of the rain and, when it had excess, this fell in streams towards the valley floor. We could see the spots where this had happened, notably at Endabash, and all the other grooves the streams had caused.

Of course, when sufficient water had descended upon the lower plateau of this great crack along the Earth, it would form lakes, such as Manyara. With no exit for this water, and since the gathering process had continued for thousands of years, Manyara had accumulated salts. The lake is still fresh enough for hippos to bask in, and for scores of animals to drink, while the tell-tale signs of soda were mainly around the water's edge. There were also surface areas where soda had collected, much as weeds or even rubbish gather on more conventional bodies of water.

A wind then blew. It rocked our basket gently, and our course became more easterly. The light began to change, shifting from dismal dark to murky grey, until it became the brightness of a normal day. The sun heated our hydrogen and soon we were rocketing upwards, as if on board an elevator, beside the dazzling whiteness of the cloud. The westerly wind which had so casually embraced us continued to carry our balloon away from the escarpment. With laughable ease, we watched the cloud recede and then the lake's eastern edge, while our shadow dot danced over the land instead of the soda. When a suitable piece of Rift Valley floor appeared ahead of us, we made full use of it, touching it once, twice and then hurtling to a halt in all its lovely dust.

In time we looked back at the Rift's features that had so nearly been the end of us. There was the great scarp, running north and south from one horizon to the other. There was the cloud, now withering, that had used the high cliff for a launching pad. Here were we on the flat valley floor, and not far away was the lake. Of course water would flow, if there was any to spare, from the higher plateau to the lower. Of course it would collect at the foot of the cliff, and then would have nowhere else to go, the valley being a slumping of the Earth's surface rather than a crevice etched over the centuries by some running river. To the west of the lake, between it and the escarpment, was a belt of forest trees taking advantage of the upper plateau's run-off. It was all so positive, and the setting sun made it clearer still. We expressed delight to each other, again and again, at the day, at our experience, at the marvel of the Rift. How amazing to have seen it as we had done. 'But Manyara means lucky,' said a Masai later on, unamazed as we told him of the flight.

The Rift lakes are all so individual. To bask in Manyara, and to emulate the hippos, would not be pleasant, save for the tremendous view. To wallow in Lake Turkana is supreme. Its water is slightly soda-soapy, and the temperature is perfect. Ethiopia's Omo river flows into Turkana, but the lake has no river outlet, hence the soapiness, although once it was connected with the White Nile over to the west. Some form of volcanic activity probably caused the barrier which isolates this vast expanse of water within the desert that surrounds it. Many fish were left stranded and today the lake abounds with them, notably the colossal Nile perch. As John Hillaby reported after walking round this jade sea, catching the perch is more weight-lifting than normal angling. They can weigh 90 kilograms apiece.

A swim in Baringo, further to the south, is also accompanied by crocodiles, but this lake is fresher, less buoyant, cooler, and less astonishing than Turkana. The more southerly Bogoria is soda-filled, like Manyara, and splendidly bath-like where the *maji moto* (or hot springs) bubble to the surface. The tap is unadjustable, and bathers must position themselves midway between the boiling effervescence and the lake's tranquil cool.

No one swims in Nakuru, which is steeped in soda and lies in a national park, while bathing in Elmenteita is currently difficult for it is steadily disappearing. Naivasha is also shrinking, and has not been so low since 1960. Its fresh water is still expansive, and provides good boating, fishing, and bird-watching, although it is dangerous when the winds get up and is no place for basking, the water being too cold. Further to the south lies that oddity which allows one to walk upon its surface, Lake Magadi. Next is Natron, flamingoes' paradise. These birds only favour soda lakes, where the algae on which they feed can thrive. They seem to operate collectively when selecting each year's soda habitat. Nakuru may be covered with the birds, or most suddenly depleted, but their preferred nesting ground along the Rift is usually Natron.

Of all the Rift's living enchantments the great flocks of flamingoes are among the most impressive. To see them striding, swimming, flying or apparently asleep, sometimes in their thousands, sometimes in their hundreds of thousands, is always a delight. They cannot be confused with any other kind of bird, and are certainly bizarre, but their way of life is most suited to some of the Rift's extraordinary soda lakes. Neither of the two flamingo species which inhabit them could survive on fresh water, although they need to drink it. They need to consume the algae and small invertebrates that flourish with such abundance in the warm, saline water, and the 900,000 flamingoes observed over several years on Lake Nakuru were thought to be taking 60 dry weight tonnes of algae every single day.

The Greater Flamingo occurs in smaller numbers and is considerably taller than the Lesser Flamingo which is the world's most abundant member of this family. Most of the Lessers are in Africa, and most of these live along the Rift. The flocks, so difficult to estimate, may be in their hundreds of thousands or may exceed a million. Why they leave one place and suddenly arrive elsewhere can often seem illogical, and the mystery of their nomadic ways is enhanced by their preference for flying at night. One day, therefore, a lake may be crowded with flamingoes;

Lake Malawi (right), *one of Africa's deepest lakes, has been isolated for several hundred thousand years, or maybe millions, and has its own collection of fish, most of which occur nowhere else.*

No fish family has become so diversified in the African lakes as the
cichlids. Whether as nest-makers or mouth-brooders (below) they seem to have
colonised every possible lake habitat. Standing $1\frac{1}{2}$ metres tall the Goliath (right)
is the largest of the herons, and is found by many of the Rift's lakes.

Whereas it may be easy to miss a frog making use of a flower (above), there is no difficulty with hippos (right), whether they are making their smug deep-throated chuckle or opening their jaws in an aggressive display.

A chain of lakes forms the Akagera swamp (overleaf) in Rwanda.
Papyrus no longer grows in Egypt, where it was once used to make paper,
but is still plentiful around many of the East African lakes.

Sitatunga (above) are aquatic antelopes, with splayed-out hooves allowing them to inhabit swamps and floating mats of vegetation as at Akagera, Rwanda. The shoebill, or whale-headed stork (right), is another swamp resident.

*In the Western Rift the Ruzizi river (above) is the principal source
of water for Lake Tanganyika, the deepest lake in Africa. The Ruzizi receives
its water from Lake Kivu (right), one of the most beautiful. Men and pelicans
compete for fish at Vitshumbi fishing village in eastern Zaire (overleaf).*

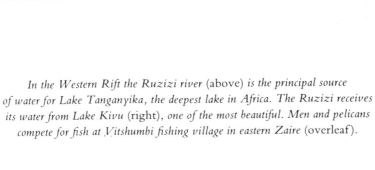

*About 80 per cent of Lake Tanganyika's fish species are endemic,
notably its cichlids seen here with catfish (above left). Examples
are* Boulengerochromis microlepis (above right), Lamprologus
brichardi (right), *and another* Lamprologus *feeding with a spiny eel*
(far right).

on the succeeding day it can be totally bereft. Meanwhile another stretch of water, perhaps 500 kilometres distant, abruptly becomes alive with birds swinging their heads from side to side as their food is filtered from the lake.

Sexual displays can be intriguing, with a few birds starting the arousal by raising their necks and crowding together. Others then join in, perhaps many thousands of others, but not the entire flock. Oddly these mass displays do not seem necessary for reproduction and, conversely, there may be mass displays without any subsequent mating. The nest is a cone of mud, with each pair raising about 30 kilograms of material to contain their single egg. For each chick the elevation is vital as the surrounding muddy water may reach the scalding temperature of 60° Centigrade, particularly at Lake Natron. Just as the birds can fly elsewhere for no apparent reason so is the period of breeding similarly erratic. In eastern Africa, although laying can be initiated between August and October or even December, there seems to be little correlation between this event and food supply or the prevailing climate. Both males and females sit on the eggs for long stretches, sometimes lasting over 24 hours. The chicks leave their mounds after eight days or so and form large groups, perhaps a quarter of a million strong. Amazingly each adult finds, recognises and feeds only its own chick, with the chicks being able to identify the particular calls of their parents.

This abundance of potential food, notably during the flamingo hatching season, might seem ideal for predation. In fact the loss due to vultures and eagles is not more than five per cent of eggs and young. The loss from mammals is almost nil, due to the unwelcoming habitat in which the flamingoes choose to live and make their nests. Fish eagles can take the adult birds, and do so notably at Lake Bogoria. One reason for the night-time nomadic excursions is thought to be a greater security from predators, and flamingoes making such a journey during the daytime have been observed flying higher than the raptors.

The extraordinarily heavy rains of 1962, remembered all over eastern

The Rift Valley lakes have always been a good source of protein, as at Lake Tanganyika (above left) or in Tanzania, where night fishing is common (below left), but carelessness with wrongful introductions of fish or oil exploitation could harm the traditional bounty.

Africa, made Natron unusable for the flamingoes. The birds therefore moved to nearby Magadi and fashioned their nests in its shallow water, a liquid much richer in soda than their usual haunts. One million nests were constructed, providing an unforgettable sight, and all went well until the chicks hatched, a single chick per nest. The adults flew off to feed each day, leaving the chicks to paddle about in the Magadi soda. Ornithologists came to admire this extraordinary scene, and were then horrified to notice the young birds staggering. Soda deposits had built up around their spindly legs until, with the more severely affected chicks, it had created most unwelcome bangles 10 centimetres or so across. All the offspring reared in nests surrounded by shallow water were affected to a greater or lesser degree. Their parents' move to Magadi had been disastrous. It seemed that tens of thousands, or possibly hundreds of thousands, would die.

The curator of birds at the Nairobi Museum quickly mobilised groups of rescuers. Schoolchildren were on holiday, and other volunteers also took part. The birds could not peck off their accumulated soda, but the humans quickly learned that one sharp tap with a hammer would crack the anklet. Each helper therefore caught a bird, positioned it conveniently, tapped one leg, then the other, and finally released the freed flamingo, trying not to dwell on the fact that a million chicks had been hatched on Magadi that year. Fortunately only 10 per cent of them suffered from the soda, but this still presented a formidable problem. In the end, although a grand total of 28,000 chicks were liberated, a greater number perished.

A flock of flamingoes is undoubtedly a joy of the Eastern Rift lakes. To observe them nesting, each pair having raised its sizeable hummock, is to be doubly privileged. Just as the Rift Valley system makes people eager to understand geology, so do the lakes inspire budding ornithologists. Standing still with a pair of field-glasses at Lake Manyara it is not difficult to list over a hundred species. To move around the 320 square-kilometre Manyara national park, and be present at all seasons, can make the final bird list nearer 400, including 13 kinds of heron or egret, three ibises, 11 ducks, four geese, 30 birds of prey, nine cuckoos and six hornbills.

Possibly even more enchanting than the profusion of one kind of species is the mix of creatures so frequently visible from one spot. On my last visit to Manyara I stayed by one of the escarpment streams leading to the lake. It was about 12 metres across, a metre deep, and the water wandered slowly between two banks from the last of the trees to the soda lake itself. Within 30 metres at most, a gorgeous amalgam was displayed – pied kingfishers, hammerkop, black-necked herons and three other sorts of heron, hippos, two kinds of pelican, cormorants, gulls, zebra, sacred ibis, wildebeest, ground hornbills, crowned cranes, the greater and lesser flamingo, spoonbills, Egyptian geese, and yellow-billed storks. The cranes strutted past the cormorants who were drying their wings. Pelicans flew in continuously, more like proud galleons than birds in flight. Flamingoes flew past, seeming wholly different creatures from the gawky ground-based members of their species. Gulls and egrets perched upon a convenient back, occasionally hurrying to the other end when the hippo gaped tremendously. There were not just a few of each kind on that single narrow stream, but hundreds: geese, storks, ibises and pelicans, with only some species, like the hammerkop and hornbill, being in short supply. I was enchanted, and did not care to move. This park's tree-climbing lions, its baboons and monkeys, its impala, bushbuck and elephants, would have to wait for another day.

Despite Manyara's excellent variety of wildlife and habitat, many people prefer Naivasha. Its escarpment is not near at hand, and so there are excellent long range views. The yellow-barked thorn trees around its edge, *xanthophloeia*, are tall and quite supreme. Bird-watchers, in particular, favour this lake. With fish eagles calling to the sky, marsh harriers almost walking through the air, lily-trotters strolling on the scantest vegetation, love-birds screeching most unlovingly, and coots by the thousand or even tens of thousand, it is an ornithologist's paradise. It is also a nothing-watcher's paradise, perfect for those who like to gaze blankly, perhaps at the papyrus fronds, or the weed-like salvinia, or merely at the vast expanse of water. Naivasha means waves, according to one Masai I asked, and winds can transform its waters almost instantaneously. Any people boating on the lake can soon find themselves in peril and occasionally the transformation can be lethal. A further peril also faces many of the lakes themselves, including beautiful Naivasha: by and large they are drying up (even if good rains in 1988 did improve the situation).

Lake Turkana, the most dramatic of the Eastern Rift lakes, is one more example of the prevailing lowering. It is 265 kilometres long and some 30 kilometres wide on average but when flying low over its waters it seems a limitless expanse. The shoreline shows many signs of its previous levels including one, less conspicuous but noted by geologists, at 75 metres above today's surface. This marks its highest level, reached some 9500 years ago, when Turkana is thought to have had connections with the Nile. Since then there have been some lesser rises, but none so high as to make a reconnection.

Practically all of Turkana's inflow of roughly one sixth of its volume per year comes from Ethiopia's Omo river. Some comes from the Turkwel, which arises basically from Mount Elgon, an old volcano well to the south-west, but this stream usually dries up for half the year. Other waterways may leap into life following a local rainstorm, and then either flow for hours or even expire before adding to the lake. All this inflow is needed to replace the water lost by evaporation, which in that hot and windy place is reckoned to be 30 centimetres a month. Evaporation has given the waters a high salt content, making bathing a buoyant pleasure. Despite the salinity, Turkana possesses many freshwater fish and is the most saline of Africa's lakes to do so. Evaporation is so considerable that the lake ought to have become much more saline that it is. Since the lake is now locked in by land, it is thought that there must be some subterranean outlet to account for this discrepancy.

There are lakes within Turkana's Central Island which are only connected intermittently with the main body of water, and are often saltier still. Even so, freshwater fish thrive and crocodiles breed in some of them, albeit the fresher ones. The tolerance of freshwater species to Turkana's saltiness is surprising, yet some scientists believe that the lake can become twice as saline before natural selection starts to bring changes, such as favouring different kinds of fish. The present concern is not salinity but the general lowering of the lake.

A fish factory was erected in the 1970s to exploit the lake's valuable source of protein, and processed 17,000 tonnes of fish in its best year. Such a hot, dry, windy area is ideal for desiccating fish, for turning them into the flattened, bone-rich boards which apparently have appeal. The catch came predominantly from the shallow waters of various inlets, such as Ferguson's Gulf where *Tilapia*, in particular, liked to feed and congregate. It is now possible to drive over that gulf, past forlorn jetties, and be amazed at the abundance of vegetation. As a result of this gulf becoming land the fish catch plunged to 200 tonnes a year by the late 1980s. The fish factory is now an empty shell, with perhaps one truck being loaded each day, a single pile of dried fish being bound into

116

packages convenient for transportation, and hardly any workers.

The world knows of the terrible rain shortage in Ethiopia during the mid-1980s, and less water undoubtedly flowed down the Omo river throughout that time. Perhaps this was sufficient to cause the lowering, or perhaps some other happening, some further faulting, some extra movement of the Rift, was also relevant. The current plan at Turkana is to step up gill-netting in the deeper portions of the lake, hopefully giving the fish factory more work to do. A fish-farm is also being built, perhaps to produce a more reliable supply, or perhaps to re-stock the lake now that so many of the breeding shallows have become dry land. It is difficult to develop such areas, and plan future operations, when all efforts are subject to the vicissitudes of nature. In so finely balanced an environment, the most well-intentioned scheme can misfire.

For example, there had been considerable starvation among the Turkana people. To help stave off such hunger in the future it was suggested that they could be persuaded to eat more fish, it being only a minor portion of their diet. As they roamed, with their donkeys and camels on the lake's western side, they were persuaded to take an interest in the fish factory (built, in the main, by Norway) near that western shore. The people then received suggestions, particularly after the gulf's transformation into land, that the fish-farm posed better prospects (it being built, in the main, by Italy). Despite the problems, and the apparent change in policy, the Turkana people did benefit from the fishing on the lake named after them, and they gained in wealth. This they invested largely in more donkeys and camels. These animals, however blamelessly, ate more of the remaining vegetation, thereby causing more semi-desert to become real desert, and making future starvation much more probable.

One virtue, in this roundabout of cause and effect, is that Ferguson's Gulf is now excellent pasture. It should have been a catching zone for fish, but the lake and the heavens decided otherwise. The cattle that were purchased while the fishing was good now graze the fishing grounds. A favoured saying up and down the African continent is that one door opens when another closes. Unfortunately, the second door may open nothing like so wide. It was quite a problem, admitted a Norwegian, teaching the inland Turkana to eat fish (just as it would be problematical instructing Norwegians to welcome donkey meat or camel with their smorgasbord). In the main the Turkana are now back again with their ancient ways and everyone is a little wiser.

The rights and wrongs of such schemes are difficult to assess, particularly when no one knows what will happen to the lake next year. Oil might be found in the area, and active prospecting is in force. The area's surveyors are based at Lodwar, a mere 65 kilometres of superlative road from the dried-up gulf. This town arose in a similar fashion to Nairobi. There was nothing of Kenya's capital until the railway engineers paused there at the turn of the century. There was nothing of Lodwar until an Indian arrived thereabouts in 1934 with donkey-loads of wares to sell. His isolated business prospered, and by the 1950s the place had grown sufficiently for Jomo Kenyatta to be incarcerated at this suitably distant spot. Colonial Britain not only removed Kenya's future first president from active politics, but helped thereby to raise his status as potential father of the nation. Lodwar's isolation had been a certain factor in the choice of prison, but it is now only nine hours by road from Kenya's capital and much less by air, there being one enormous runway running straight into town. If oil is found near Lodwar the place will change even more dramatically than in its first half-century.

At present the guests who arrive at the Ferguson's Gulf hotel can savour the virtues and misfortunes of this Rift lake. Instead of reaching the thatched oasis via motorboats, and across a gulf teeming visibly with fish, they bounce to it in a motor vehicle rich only with talk about former fishing days. The tilapia have yielded to cattle, to a pastoral luxuriance in that hot and arid land. The visitors can still enjoy the warm, soda-tasting wind that comes always from the east, and watch the egrets walking by the shore. They can ponder on the lake itself, how it has risen and fallen in its own good time, and has cared nothing for the humans making use of it. People have been living here, or so the fossil record says, throughout much of their hominid history. They too must have wondered about the lake of their ancestors, as it rose unaccountably or fell no less dramatically, and they certainly had to make do with a changing situation. That was their ability, as the Rift flexed its brand of muscles, causing drought here, abundance there, while the environment was rearranged. It is easy on this wilderness of a lake to feel at one with our ancestors who savoured the same stickiness and warmth. It is an excellent place, however variable, at which to spend some time.

·V·

THE
WESTERN
RIFT

—

The lakes along the eastern portion of the Rift each have their individual fascinations, but the Western Rift is possibly even more extraordinary. Its northern end begins more or less where the White Nile hurries in and out of Albert/Mobutu Sese Seko. Southwards, and linked by the Semliki river, are lakes George and Edward. Next in line comes Lake Kivu, and then Lake Tanganyika, the world's second deepest lake, beaten only by Lake Baikal in the Soviet Union. Lake Malawi, further to the south, is Africa's second largest lake, after Victoria. Geologically it is more of a continuation of the Eastern Rift, but visually it appears to partner Lake Tanganyika and is therefore detailed in this chapter. North of Malawi is Lake Rukwa, and south are lakes Malombe and Chilwa. Finally, although not a lake at all but still a substantial spread of water, notably near its mouth, lies the Zambesi river, which brings this magnificent chain to an end. The eastern lakes are better known, better visited, more famous, and closer to large centres of population than their western counterparts. They have their oddities, such as Lake Magadi, but it is arguable that more strangeness occurs over to the west, for example, at Lake Tanganyika.

Europeans saw this lake in 1858. Richard Burton and John Hanning Speke, obsessed like all east African explorers of the time with the River Nile, were first to encounter this huge body of water but were ultimately disappointed with their find. It did not flow towards the Nile – indeed it did not seem to flow anywhere – and was therefore unrewarding. Speke picked up some mollusc shells, resembling marine shells, and thereby started a controversy that is not settled yet. Did the lake ever link with the sea, either westwards via Zaire or eastwards to the Indian Ocean? Although high escarpments prevent significant inflow, Tanganyika currently receives water from Lake Kivu via the Ruzizi river, from the Malagarasi which drains an eastern area, and from several minor rivers that come and go in standard African fashion. Sometimes the lake overflows to send water westwards down the Lukuga river, but this is an intermittent affair. Most water loss is by evaporation. However, its salinity is not as high as one might expect if evaporation has always been its principal source of water loss. Its salinity, for example, is five times lower than that of Lake Turkana. Possibly the Lukuga outlet was more consistent, and more important, in former times.

Lake Tanganyika is roughly 650 kilometres long by 50 kilometres wide, giving it an area approaching 33,000 square kilometres (which is greater than Belgium). Its surface at 760 metres above sea level is not high in African terms, but it sinks to a maximum depth of 1470 metres and so part of the lake bottom is nearly as far below sea-level as its surface is above it. This great depth gives it a tremendous volume, roughly 19,200 cubic kilometres. Most of this is quite unsuitable for life, being so rich in hydrogen sulphide and so deficient in oxygen that nothing can survive. Only within 200 metres or so of the surface does the water cease to be 'dead', as the lower levels are frequently and accurately described. The depth of water available for life is even shallower at the lake's northern end, being no more than 45 metres. Underneath that upper layer is not so much death as an absence of life, and the deepest trawl ever to catch a fish was working 240 metres down, near the lake's southern end. The world's oceans, by contrast, contain life all the way to the bottom, even in the deepest areas 11,000 metres below the surface.

Should some form of upwelling occur to break the normal separation of levels in Tanganyika, there would be tremendous mortality, with fish in particular, being asphyxiated by the lack of oxygen. For scientists, the undisturbed depths have considerable appeal because they contain what is known as relict water. They are time capsules from previous ages, and can therefore provide undisturbed information about the past. Of the world's lakes only three contain significant amounts of relict water, namely Kivu, Tanganyika and Malawi, and all are in the African Rift.

An equally bizarre feature of Tanganyika's water is its consistent temperature at any level. From the surface to the depths there is a drop of only 3° Centigrade. In the middle of the lake, and below the immediately superficial layer, the temperature is 26.5° Centigrade. About 900 metres below that spot it is 23.3°, and may even warm a little near the actual bottom. It is not 'fully understood', to use the beloved phrase of science, why Lake Tanganyika is so stable in its layering, its temperature, its lack of oxygen and abundance of minerals in the lower depths, and this riddle may never be solved. Possible factors include the type and quantity of incoming river water, the amount of sunshine and of wind, and even the planet's rotation. Anyone who has been on this lake during one of its violent storms, most of which occur in the rainy season, may be astonished that so little mixing is taking place deep below the surface. The upper layer is in turmoil, with frightening waves up to six metres high, and the whole lake seems disturbed.

Tanganyika is thought to have been created in the Pliocene period. Its antiquity is important, but its degree of isolation is much more intriguing. Tanganyika has experienced more or less the same conditions for a very long time, probably longer than any other African lake. Its depth has gone up and down, and for much of its history it was about 550 metres below its current level. However, it is thought to have had a stability not enjoyed by other lakes which, in general, have either been dry from time to time or were once united in some bigger sea. In recent years the lake has experienced only modest change. During 1962-63, when eastern Africa was drowned by the tremendous rains which forced the flamingoes from Natron to Magadi, huge Lake Tanganyika rose some six metres above its current level. Since then, it has been dropping by about 45 centimetres a year. Even when the Lukuga outlet is functioning, sending lake water to join Zaire's Lualaba river, at least 90 per cent of the lake's water loss is due to evaporation.

Lake Tanganyika is a dangerous place to be caught during one of its many storms, partly because the high waves can sink small craft, and partly because so much of the shoreline provides no relief for anyone seeking shelter. The great escarpments, even more impressive on the lake's west side than the east, have gradients of 30-40° and continue on downwards without any hesitation. The lake's formidable depth is impossible to visualise. However, comparisons between this lake's basic parameters and those of other lakes help to demonstrate the fact. Victoria,

Africa's largest lake, has an area twice that of Tanganyika, but its volume is only one-seventh of the less extensive, but far deeper, lake. Malawi is closest in size, being almost as long as Tanganyika and containing 44 per cent as much water. Lake Turkana is the largest lake of the Eastern Rift, and looks impressively huge even from aircraft at 30,000 feet, but its area is one-third that of Tanganyika and its volume is one-eightieth, a major difference being that Turkana is nowhere more than 120 metres deep.

The depth and volume of Tanganyika has enabled it to survive, reasonably intact, during times of climatic change. If Turkana's rivers switched off their input this lake would be dry, perhaps within a dozen years. If Tanganyika were to be similarly starved of inflow a dozen *centuries* might pass before its water had vanished. This relative stability has been of considerable influence, permitting the residents to evolve in their own time, protected from drastic change to their environment. Some of the eastern string of lakes have altered dramatically, even within one human's span. I have seen Lake Elmenteita transformed from dry to extensive to dry again within no more than 30 years. The huge Ferguson's Gulf, on Turkana's western shore, seems to have dried up overnight, so swift was the change from an excellent spawning and feeding ground for fish to a dusty pastureland more suitable for cattle. It would seem that Tanganyika and, to a slightly lesser extent, Malawi have never experienced such a revolution.

As a result, these large, stable, deep lakes can be viewed as islands, isolated from other areas but persisting with their trapped and evolving forms of life. Indeed, they are far more cut off than many islands. Britain, for example, has been an island for only a few thousand years, and during this time all sorts of alien species have been introduced, including one quarter of its present-day mammals. In contrast, Lake Tanganyika has survived in isolation for millions of years, permitting new species to arise or considerable modification of existing forms. As a result a high proportion of the lake's animals are endemic. Their ancestors must have been imported from the rivers which drained that area of Africa before the rifting, but much of the fauna has changed so radically that it is now extremely dissimilar to the original river stock. In Lake Tanganyika the majority of its fish species are endemic and one particular fish family, the cichlids, has diversified to an extreme extent. Whereas 57 per cent of the non-cichlid species are peculiar to this lake, a phenomenal 98 per cent of cichlids are endemic.

A similar story exists both for Victoria and Malawi, but in Tanganyika the deviation has been widest. Possibly the lake's waters have been more to the fauna's liking, or the lake has been isolated for longer, or some other variable has favoured evolutionary change. Whatever the reason, the lake has become both a museum of relict water and an active zone of alteration. A genus is a division within a family, and usually contains several, or even many, species. Excluding its numerous cichlids, Lake Tanganyika has eight endemic fish genera, whereas Victoria has one and Malawi none.

Other animals have also taken advantage of Tanganyika's insularity. All seven of the lake's crabs are endemic as are five of the 13 bivalve molluscs, 37 of the 60 gastropod molluscs, and 11 of the 33 copepod crustaceans, which provide food for so many of the fish. Higher up the animal kingdom, there are two snake species unique to the lake, a colubrid and a cobra. There is nothing strange about the lake's hippos or crocodiles, save that its Nile crocodiles seem fonder of taking humans by the Ruzizi than is normally the case in Africa. Pierre Brichard, who has spent much of his life around Lake Tanganyika, has written that in this river's estuary 'many people are seized, crushed and eaten every year by crocodiles, to such a point where half of the crocodiles killed by hunters have human remains or artifacts in them'. As for the hippos, who allegedly kill more people than any other African animal, Brichard recommends that skin-divers make immediately for deeper water in case of a sudden encounter. He thinks it unlikely that a hippo could go very deep but makes the disconcerting point that these animals, so apparently bulky and ungainly, can travel faster underwater than the sleekest fin-footed human.

As every canoeist or anyone who has descended a river knows, the sudden arrival at a lake is most dramatic. The hurrying turbulence turns to tranquillity, the shallow waters deepen, and every other aspect seems to change: the clarity of the water, the kind of vegetation, the shore-lines which replace the banks. How much more astonishing this must be for any kind of fish. There are new dangers from different kinds of predator. There are brand new habitats to be explored, and colonised. There is depth, and no need to maintain station against a tireless current. There is different food, unable to grow in rivers, but amply suited to a lake. Should the huge body of water have no outlet, as (usually) with Tanganyika, there will also be increased salinity.

Not every river fish can adjust to such conditions. Some fail, but others flourish and then radiate to take full advantage of the novel scene. Clearly at Tanganyika the cichlids were most able to change from riverine to lacustrine, from torrent to lake. In contrast, fish species in the Malagarasi river, which flows into Tanganyika from the east, are not represented in the lake but are found in the Zaire basin over to the west. Presumably, the Malagarasi was once an upper tributary of the Zaire and its river fish did not appreciate the lake. As the vastness of Tanganyika did not arise overnight some of these fish may have survived in its waters, perhaps for a million years or so, before dying out.

It is understandable that the new environment did not suit all river species. Today, the lake has a pH of 9.0, making it much more alkaline than standard river water. It has large quantities of salts in solution, such as sodium and magnesium carbonate. The lake is warmer than its rivers, except where the Malagarasi flows through a long delta, permitting its waters to be heated even above the lake's temperature. From about six metres down calcium carbonate covers everything in its crystalline form, calcite. This fuses pebbles, rocks and so forth, thus altering the kinds of bottom habitat. Most of the shoreline is extremely steep, making the lake's periphery more like the sharp drop of a coral reef. Elsewhere, waves hammering at the cliffs have caused profusions of fallen rocks, as well as accumulated sand and mud. In many ways Tanganyika is more like a sea, with increased salinity, varying kinds of shore, and considerable surface turbulence.

Two-thirds of this lake's fish species belong to the perch-like cichlid family, the remaining one-third being spread among 19 other families. Cichlids, native to South and Central America, Africa, Syria and coastal India, have flowered magnificently in the post-Rift African lakes. In Tanganyika there are now (at least) 139 species and five sub-species representing 42 genera. The cichlids seem to have adapted themselves to every situation the lake offers, 20 per cent of the species living in open water, 40 per cent living along the rocky shores, 20 per cent linked to the sandy or muddy bottoms, and the remainder less precise.

Most fish reproduce by squandering large numbers of sperm and eggs and then disregarding their potential progeny, which may or may not survive as viable offspring. The cichlids do not employ casual methods. They either lay their eggs in some form of nest, almost bird fashion, and then care for the growing young, or the parents themselves provide the nest by sheltering the young within their mouths. This fish family does exist in large numbers in other parts of the world, and yet nowhere else, as Brichard phrases it, 'do they appear to have brought the techniques inherited from their ancestors to such diversity and fine tuning'.

Boulengerochromis microlepis is the only species of its genus, and also has the distinction of being the largest cichlid. It can achieve a length of 75 centimetres, with heavier females weighing about 4.5 kilograms. This cichlid's nest is also huge, being 3.5 metres across and just under a metre deep in the sand. It lays 10,000 or more eggs at a time in this giant saucer of a depression and virtually all are fertilised. When the young hatch after a couple of days, the central and deepest part of the nest becomes a squirming, writhing maelstrom of small fry. Both parents stay in the vicinity, ready to defend their young. The adults have no particular predators, as the crocodiles stay close inshore. Nile perch feed mainly on the many smaller fish, and large predatory catfish stay near or in their caves. However, many smaller piscivores are quick to pounce on the young within the saucer-shaped nests in the clear, well oxygenated shallow waters. The parents do their best to protect their many thousand offspring, particularly during the most vulnerable first five days before the fry can swim, but nevertheless losses are considerable. The many thousands which hatch are cut down to a few hundred before the survivors leave their nest and all parental protection.

Many other cichlid species lay a smaller cluster of eggs, and spend much energy in fanning and aerating the developing young. *Boulengerochromis* adults do no fanning, having enough to do defending their offspring. Instead, the fry aerate themselves. As soon as the minute larvae are able to wriggle their tails they do so in such close unison that they seem to act as a single entity. There is also a circulation from bottom to top, with those in most danger of being oxygen-deprived struggling for the surface. As the fry grow larger they maintain this communal approach, resembling a single ball rather than a profusion of individuals. They even feed in this manner, with those on the circumference eating before their prime position is usurped by others emerging from the centre. The parents still watch carefully over this whirligig of youngsters, behaving like none other in the lake.

When the offspring are about a centimetre long, they have exhausted their yolk sacs and become free swimmers. While still suffering extensively from predation, they set about revenge, and eat steadfastly and voraciously. Within a few months they are 10 centimetres long. As they grow they gradually become safer from predators, and are increasingly able to prey on others. Initially they hunt in schools of between 100 and 5000 individuals, consuming almost anything they can catch. Later the numbers are reduced to several dozen in each group and the fish roam more widely, leaving the shallow waters for the deeps and eating only fish, preferably those a quarter or less of their own size. They are thought to reach sexual maturity at 30 months when they each weigh about half a kilogram. They then pair up and become, along with others of their species, the only Tanganyika fish to make their nests out on the open sand, where no protection exists save for that the parents can provide.

The *Boulengerochromis* way of life and breeding system is more suited to a lake than a river, however much this largest of the cichlids did originate from a river system. Such a huge crater of a nest could never have been constructed in a stream, and the ravening appetites of predatory schools of 5000 youngsters need large food sources to sustain them, such as those present in a lake. Among its many successful modifications *Boulengerochromis* did make one mistake. It should never have permitted itself to be both large *and* extremely tasty, attributes which may yet see it replace the popular *Tilapia* in fish farms.

Another cichlid, *Cyathopharynx furcifer*, fashions quite a different sort of nest, often referred to as a penthouse. The male builds the considerable structure entirely by himself. First he chooses a suitably high and attractive boulder. Then, taking one mouthful of sand after another, he painstakingly constructs a nest about 30 centimetres in diameter on the summit of the rock. This mighty labour is achieved by a fish only 18 centimetres in length. The mystery in all this activity is that *Cyathopharynx furcifer* is a mouth-brooder and so does not need to fashion a home for its developing young, although the penthouse does bring both the male and his nuptial statement nearer to the females who tend to circle in higher waters. In any event, having created his couch with such exertion, the male swims upwards and entices a ripe female down to his mountain top. When she enters the nest he changes colour from his basic grey to an iridescent blue. If the pair is undisturbed, the colour continues and stimulates the female into laying eggs. She does not produce the thousands squandered by some species, but emits one or two at a time. Simultaneously the male ejects sperm. Once her modest number of eggs has been laid and fertilised, the female takes the developing cells in her mouth and leaves the nest, the thing having served no more than a mating couch.

126

After the female has departed, the male will attempt to entice another mate. If he is again successful, and she favours his penthouse, the mating procedure will begin again. As each female is mated, she continues to keep the developing eggs in her mouth. Like all fish eggs, they first develop into larvae which gradually absorb their yolk-sacs. When they turn into fry, she still protects them in this buccal fashion, liberating them from time to time to feed and taking them back when danger threatens. Eventually they are allowed to go their own way. In due course, half of the survivors will collect suitably sized grains of sand, take them mouthful by mouthful to some rocky prominence, and fashion a honeymoon home.

The reproductive strategies developed by the cichlids in Africa's great lakes are so varied that it would seem every possible behavioural trick has been employed. Unlike the polygamous penthouse dweller, and unlike the nest-protecting *Boulengerochromis*, the beautiful, rock-living *Julidochromis marlieri* pairs for life. This species also produces and cares for a minute number of offspring at a time. Each female produces only a dozen or so eggs. Each small family is raised within particularly difficult crevices in the rocks, thus reducing the chances of attack by predators. Having opted for a minute quantity of young, this species tries to ensure that all survive, giving them a degree of attention which seems in harmony with the lifelong monogamous relationship. A further peculiarity is that the young are raised sequentially. Most fish indulge in one frantic flurry, liberating sperm and eggs in a frenzy of reproduction, and then suspending such activity until the following season. *Julidochromis*'s brood is more like a human family. Each chamber in the rocks is the home of an adult pair, and may also contain eggs, larvae, small fry about one centimetre in size and larger fry, almost three times as big.

This succession of miniature spawnings, leading to a successful co-habitation by different age groups, must have advantages. By liberating thousands of young simultaneously most fish species create a surfeit for predators which allows a few to survive. *Julidochromis*'s technique of slipping its young out one by one carries the risk that predators could pick off the juveniles as they emerge. However, it has the advantage that the larger fry leave the parental home when they are ready to do so, and not in some Gadarene tumult certain to attract predators. Also it gives each youngster a better chance of finding a suitable rocky niche in which to secrete itself than if all its siblings were simultaneously competing for sanctuary. The pair-bonded parents cease to care for the offspring once the young leave the chamber, but have ensured a far greater chance of individual survival than most other fish manage to achieve.

The strategy of each species is carefully suited to its particular habitat. *Boulengerochromis* lives above and upon the open sand, *Cyathopharynx* prefers boulders, and *Julidochromis* requires a profusion of rocks, from which it can dart to feed. These three fish are all of different genera, but even closely related species can vary their behaviour and so exploit the different niches available to them. The genus *Lamprologus* has diversified so much that it has become the most important cichlid group in Tanganyika. Its 35 species make up a quarter of all the kinds of cichlid to which the lake is host. The species of *Lamprologus* that feeds on plankton about a metre above the bottom is particularly interesting. *Lamprologus brichardi* is a tiny fish, only about six centimetres long. Fine filaments extend from the ends of its fins, notably the tail and dorsal fins, which makes the species desirable to those who keep fish in aquaria. These ornaments and its lack of streamlining suggest correctly that *brichardi* is incapable of speed, but it has no need of haste in catching the drifting plankton, and never strays far from shelter in case danger looms.

A further distinction is that this cichlid lives in a school, sometimes 100,000 strong. Its school is different from the normal variety, in which a crowd of individuals move through the water in a unified fashion. Such groups have apparently a single thought, like flocks of birds twisting and wheeling in total symmetry. *Lamprologus brichardi* behaves more like a collection of human beings gathered in a town, displaying unity but also a collective disparity. Its school hardly moves, often feeding for years over the same rubble-strewn patch of lake bottom. Some ripe adults pair up, while others form a different co-habiting arrangement, such as one male acquiring a group of females. Each pair or group selects a suitable recess and then spawns within it. The nests are often close together, and the adults in one locality may join forces to protect the developing fry.

Finally, as a most remarkable form of fish behaviour (although often witnessed in human families), the older fry help to protect the younger ones as soon as they are able. This species of *Lamprologus* therefore only appears to form a school. On closer inspection it is a crowd of individual communities, a collection bonded by their close proximity. Watch a rally of human beings, of assembled people, and then observe them going home, each small group returning to its chosen crevice. *Lamprologus brichardi* is only one of many cichlid species living in the area, but what

*A satellite view of the Sinai Peninsula (above) showing the Red Sea's
northerly enclaves, the gulfs of Suez and Aqaba.*

The Rift system near the Sea of Galilee (left and below), *and the Yehuda Desert and the Dead Sea* (bottom).

*The Egyptian Desert (above) close to the Red Sea, with old and
well-eroded Rift Valley escarpments. The Arabian oryx (right) originated from an
African group, but became isolated on the Red Sea's eastern side.*

Ospreys normally nest above ground, usually in trees, but around the Red Sea they have to nest at ground level (above and top right). *The white-eyed gull, endemic to the area, nests at the height of summer and the chicks must be protected from the fierce heat* (centre and bottom right).

Anthias (left) *live throughout the Indo-Pacific Ocean, but are particularly abundant in the Red Sea where the upwelling of water can lead to ample supplies of plankton. Goat-fish (below) and butterfly fish (bottom) are also common around the coral reefs.*

*Scorpion fish (left), brilliantly camouflaged, venomous and predatory,
are not endemic to the Red Sea. A trigger fish (below left), shows its formidable
teeth as it guards its nest of young, and a parrot fish (below right) is attended
by a cleaner wrasse.*

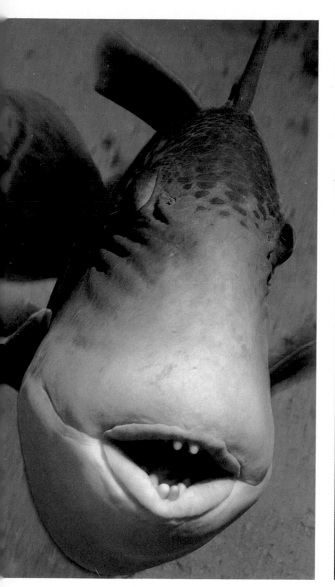

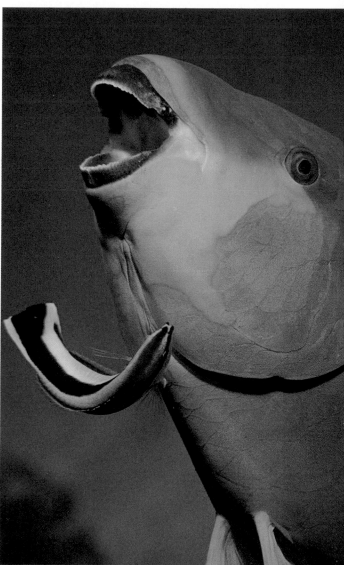

*A clown fish (above), endemic to the Red Sea, secretes a fluid that
permits it to find security within an anemone's tentacles. The frog fish (right)
is a camouflaged angler with an attractive lure on the top of its head.*

The Red Sea contains the most northerly coral reefs in the world
(left). *Although many fish are only found in its waters there are those, such as
the white-tip shark (below), that are able to swim in and out of this arm
of the Indian Ocean.*

unites them all is the extraordinary neatness of the cichlid system, the fine-tuned harmony which this single family has succeeded in creating within a single lake.

No wonder alarm has been expressed over current plans for finding oil beneath the waters of Tanganyika. If oil is located in significant quantity an accident, perhaps of the kind that occurred at the Ixtoc well in the Gulf of Mexico, cannot be discounted. This well blew its top in the late 1970s and belched a particularly nasty form of pollution into the sea for many months. Even a particle of that tragedy could destroy the precarious balance of Tanganyika. It is not an open sea, irrigated by the waters of the world, but a closed system, and its inmates have adapted magnificently to the only kind of world they know. To change that world, even minutely, could prove a catastrophe. Lake Tanganyika is a true relic from the past, existing most satisfactorily within the present; but one major oil spill, or a surfacing of dead, deoxygenated water, or any awful alteration, could change that excellence instantly.

The mere transference of a fish can cause disaster. The non-Rift Lake Victoria has not been as stable as Lake Tanganyika, but there used to be a certain harmony in its waters between predators and prey. Then some 30 years ago, *Lates*, the huge and famous Nile perch, was introduced to the lake. This fish had previously existed in Victoria, as shown by the fossil record, but had died out. No one is sure why it vanished but one theory, proposed by Humphry Greenwood of Britain's Natural History Museum, suggests that local volcanic activity, perhaps from Mfumbira, made the waters unacceptable to this large fish. It is known to be more susceptible to raised levels of carbon dioxide than, for example, the cichlids, and the deoxygenation caused by falling ash might have been selectively lethal. At all events the Nile perch perished, and other species continued to survive, notably the cichlids. Then came the modern introduction.

The Nile perch is a piscivore, and it set about eating the resident fish. It performed so well that certain species in Victoria declined dramatically

This brilliantly coloured soft coral, Dendronephthya, *and a brittle star (left), are also found in the Red Sea.*

in number and would appear to be in danger of extinction. The local fishermen, who had been steadily grateful for the lake's traditional bounty, were appalled by this disappearance. They might have been grateful for the new predator's considerable size but even its bulk was an impediment; it did not dry in the sun as effectively as the smaller prey it had busily consumed. The smaller fish, more abundant, tastier, and most convenient, had much to be said for them. A lion, as it were, had replaced the gazelles and no one liked the change. The greatest sufferers have been a cichlid group called the haplochromines which had done particularly well in Victoria – until the perch arrived.

The gloom expressed when the success of *Lates* was becoming blatant has now lightened a little. The huge perch is present in only small numbers in certain areas. Endemic species thought to have become extinct have been rediscovered. At one time, researchers failed to find a single cichlid at some locations, such as Kavirondo Gulf. Other icthyologists, horrified by this news, have since found them in their thousands. Best of all, for those who welcome rough justice, the hunger of the Nile perch is being assuaged by the consumption of smaller Nile perch. A sort of balance will surely emerge as the new predator, far larger than any existing in the lake, becomes controlled in number by the scarcity of food. What is not known is how far this new balance will have shifted from the previous equilibrium. Will some species have become extinct? Will the fishing industry survive or will a valuable source of protein do little more than maintain a most unwelcome crowd of perch? Or will Victoria be able to adapt to this man-made change as all the lakes have done so successfully with all the natural vicissitudes to have come their way? Finally, when will humans realise that their tinkering, blessed with good intentions, generally does harm?

Of the greatest African lakes, Tanganyika is the deepest and Victoria is the largest in area, with Malawi lying somewhere in between. Malawi's area is slightly less than half of Victoria's vast spread, but its maximum depth is almost ten times that of Victoria and its volume is three times greater. Its greatest depth is one half that of Tanganyika's and its volume less than half of Africa's greatest freshwater reservoir. However, its area is similar and Lake Malawi has more parallels with Tanganyika. These two largest lakes of the Great Rift are, by Africa's standards, extremely uniform.

At 460 metres above sea level Lake Malawi is lower than Tanganyika, which again is lower than Lake Kivu further to the north. Malawi is younger, but sufficiently old to have produced an assortment of endemic fauna almost as astonishing as that of Tanganyika. It actually contains 245 fish species, as against Tanganyika's 214, but far fewer families are represented. (These species totals will certainly increase as and when more work is completed concerning the populations of Africa's lakes.) Once more it is the Cichlidae that have enjoyed the most interesting development, resulting in 190 species of which almost all are endemic, and showing themselves astonishingly able to take advantage of the new, non-river, isolated and relatively stable conditions.

David Livingstone, so much more a geographer than a missionary, put Lake Malawi on the map for the world beyond Africa. He explored the lake in the early 1860s together with John Kirk who, when based in Zanzibar, became a one-man launching pad for British travellers wishing to explore eastern Africa. Just as Speke collected molluscs from Tanganyika, and thereby stimulated scientific interest in the lake, so Kirk collected fish from Malawi, thus showing zoologists in Britain that it too was no ordinary lake. It has one outlet, the Shire river to the Zambesi, but this is intermittent (and was closed as recently as 1915-35). Lake Malawi possesses Zambesi fish, but only those occurring in the upper section of that great river. The Shire flows into the lower Zambesi via the Murchison rapids, which descend 280 metres in 80 kilometres and provide an effective barrier against fish that might otherwise travel upstream. Plainly a much stronger biological link with the upper Zambesi existed in the past.

The Rift's lakes were first encountered by Europeans only in the latter half of the 19th century. Scientific work did not begin on them until the 1920s and only in a determined fashion after World War Two. Therefore many of their secrets have yet to be explored. Even their ages are altering as more facts come to light. Initial estimates put Victoria at about 750,000 years old, Malawi slightly older, perhaps a million years, and Tanganyika oldest of them all, between three and six million years. The recent discovery of some lake sediments beneath Lake Tanganyika over 900 metres deep has caused a certain revision of its age. Perhaps those waters have existed since the Lower Miocene, say for 20 million years. Since Malawi is such an apparent partner to Tanganyika, it too may be as old. In that expanded length of time there could have been some marine connection, a temporary transgression from the sea, which could help to explain the great array of marine-like molluscs living in the deepest, oldest lake. For a time the 'thalassoid' argument reigned, which linked them with the sea, but later work seemed to prove their relationship with freshwater forms found elsewhere on the continent. More recent theories suggest that the molluscs appear marine-like because the adaptations which equipped them for life in the lake's salinity are similar to those required for the saltiness of the oceans. In any case the double chain of lakes, such a remarkable feature of the Rift, will surely solve some of these questions, and pose others, in the years that lie ahead. However they will be unable do so if they are disrupted, if their remarkable isolation is to be destroyed by some careless intervention. Humans have done terrible things to some of the world's islands. There is scope for similar damage with the world's most astonishing group of lakes.

· VI ·
RED SEA ENCLAVE

A crucial part of the Afro-Arabian Rift, in that it separates Africa from Arabia, is the Red Sea. It is hardly a lake, and is even connected to the world's oceans, but it has a degree of isolation that makes some of its properties more like those of a lake than a piece of the open sea. In any case it is a direct link between the rifting visible in Jordan and Israel and that in the Afar depression of north-eastern Ethiopia. Although the Red Sea is clearly connected with the Gulf of Aden, and therefore with the Indian Ocean, the straits of Bab al Mandab (the 'gate of tears' in Arabic) are only about 30 kilometres across and some 100 metres deep. This constrains access between the Sea and the Gulf. However, the very narrowness of the straits makes this a popular crossing point for birds wishing to reach Africa from Asia, and vice versa. Birds of prey from Asia, such as steppe eagles and steppe buzzards, for example, fly across these straits in their hundreds of thousands every year, preferring to avoid travelling over water wherever possible.

More than once throughout its history the Red Sea has been disconnected from its parent waters, and more than once the waters have come surging in again. Whether actually united with the outside world, or temporarily severed from it, the enclave has provided suitable conditions for its own private evolution. Its coral reefs, different from anything existing outside, have been particularly effective in permitting new species to evolve. These new forms have remained within the Red

Sea, the Bab al Mandab acting as a barrier, and the enclave now contains a small proportion of endemic forms. Although the Suez canal was cut at the Red Sea's northern end in 1869 this has not greatly affected the Sea's isolation, with such flow as does exist tending towards the Mediterranean.

Like many of the lakes in the Western Rift, which are extremely deep and would be inundated with seawater were there a suitable passage, the Red Sea's maximum depth is over a kilometre and a half. Its width is increasing at a rate of about 2.5 centimetres a year, proving that the rifting process, tearing Africa from Asia, still continues. This rate may appear extremely slow but, in geological terms, the haste is considerable. It represents an increase of 25 kilometres in a mere one million years.

The Red Sea is a paradise for the diver. Its coral reefs are particularly satisfactory places, and the swimmer can coast along the top of them, glorying in their shapes, colours, forms of life, styles of life. There is always so much to see; shoal fish acting in gorgeous unison, twisting as one, approaching, retreating, and either blatantly visible or suddenly no longer there. A big grouper, less agile, and certainly less beautiful, can have great personality, almost inventing games and undoubtedly making friends. With luck one can spot cleaner-fish and, with even greater luck, a turtle.

To be captivated by this close-up study is to render oneself liable to sudden vertigo. The edge of the reef can appear unannounced, unexpected and quite vertical. From being 30 centimetres or so beneath one's belly, the coral is abruptly replaced by a blackness, apparently descending for ever. There is no reason why a swimmer should also plummet downwards, but it does not seem that way. There can be a hurried scrambling back towards the coral, a careless haste reprimanded by the first available urchin spine, an undignified retreat to the security of having something, anything, below one's flippered feet. Then the day resumes its enchantment. Perhaps the fact of being stone-deaf, with nothing but turbulence sounding in one's ears, can help abolish normal concerns, such as the sun beating down from up above. It was only after I had emerged after one long swim that I realised how much the sun had roasted me. For two days I stayed room-bound, searing hot and goose-pimpled simultaneously, resenting every twitching of my skin; but a portion of my misery lay in knowing the reef was still out there, with all its delights on show, and me not seeing them.

To explore with a bottle of air upon one's back is, of course, much better. It is usually possible to glide in the shallow water above the coral, and then swoop down into gaps to see what else the reef has in store. Sometimes masked butterfly fish parade, always in pairs, each with black patches around their eyes. Surgeon fish need to be approached warily; their scalpel-like defences, set slightly forward of the tail, can cut attackers (or passing divers) who come too near the thrashing tail. There are also the colourful slug-like nudibranchs, such as the Spanish dancers which can grow to 30 centimetres long. The clown fish are no less brightly coloured, and stay safe within anemones. At times all the colours seem excessive, with even algae bright red or yellow from time to time, and the corals almost any shade from their basic lightish green.

Television films can imply that there are animals every way one turns. In fact, which is better still, the creatures have to be sought out, likely spots have to be investigated, and rewards are therefore doubly treasured. The crocodile fish are splendidly camouflaged as they wait for some luckless passer-by. A stone-fish remains obscure even after being disturbed and having given its presence away by movement. It is never aggressive, its sting is entirely defensive, but neither fact lessens the pain should that dorsal spine be forced to find its mark. A crown-of-thorns starfish may be nearby, looking precisely like its eponym. The *Acropora* coral resembles heather, except that heather does not break one-tenth so easily.

To swim at night is to see yet another world. The corals become alive, with their little white polyps alert for food. Small shrimps and crabs venture forth when the big creatures of daytime are somnolent if not asleep. Torches are a necessary precaution, as is some form of fluorescent marker back at the boat, but switching off the torches reveals the true magic of the nocturnal reef. Illuminated fish pass by like ocean liners, with portholes all alight. The moon lights up the scene and even Venus is sometimes visible from a couple of metres or so below the surface stillness. The arrival of day transforms the entire street scene of the reef, with the cleaners – shrimp and fish – busy at their task, the clams sucking and blowing, the lurkers lurking, and the parrot fish, the trigger fish, the scorpion fish, and maybe a shoal of goat-fish all beautifully busy at their different activities. To be at one, temporarily, with the elusive angel-fish increases the sense of privilege. Its blue face, purple body and yellowness towards the tail seem much like perfection. How odd that most people knew so little of this other world until a very short time ago. It was like not knowing of a different kind of land, such as mountains for those who live upon the flat, or flatness for those whose paths are always steep.

The Red Sea is unique in so many ways, not least in its coral. Its waters contain the most northerly reef-building corals anywhere in the world. As with the deep lakes of the Western Rift, the water temperature is only initially reduced with depth. Thereafter it stays fairly constant, and actually warms towards the bottom. At its surface the sea is subject to some of the most severe conditions this planet has to offer. Fierce solar radiation has helped to create major deserts on either side, and the hot dry winds coming from those deserts aid the extreme evaporation which, on the nearby Aswan reservoir, causes water losses of over 3.5 metres a year.

The Red Sea is not only an enclave from Aden's gulf and the greater waters of the Indian Ocean; it also possesses two enclaves of its own. One leads north via the Gulf of Aqaba to Eilat, and the other leads north-west via the Gulf of Suez to Suez itself. Just as the Red Sea is isolated, to some degree, from the oceanic waters so are these two gulfs, although connected with the Red Sea, somewhat separate entities. The one running to Aqaba is more on the line of the Rift as it continues, via the Wadi Araba, first to the Dead Sea, then up the Jordan valley to the Sea of Galilee and eventually up the Beqaa valley towards Homs and Aleppo.

The other portion of this divide, the Suez half, used to be more significant. During the Miocene, when so much of the rifting was occurring, the Red Sea was a part of the Mediterranean rather than the Indian Ocean. There was no connection at the southern end, and all the fossils associated with this period are of Mediterranean forms. The northern link was then unreliable, being only intermittently open, and the Red Sea suffered in consequence. On occasion it would dry completely and become no more than a salt pan, causing the inhabitants to perish. Even if the waters did not vanish completely, the considerable evaporation would concentrate their salinity devastatingly. Then some five million years ago, the Suez area became a barrier once more and the southern end opened. For the first time the Red Sea could be populated by animals from the Indian Ocean. The species living in the Red Sea at this time may have been killed off by the southern invaders, or died during some intermediate phase when perhaps both doors were closed. Whatever their fate, the single certainty is that no descendants survive in the Red Sea today of what are known as Palaeo-Mediterranean fauna.

Life in the sea between Africa and Asia has not been tranquil since that dramatic alteration in the early Pliocene from a northern to a southern mixing. Throughout the past two million years, as the ice ages ebbed and flowed, there has been a worldwide raising and lowering of oceanic water levels. The Red Sea has not been immune, particularly when the ice age was at its most extreme, lowering the oceanic surfaces by 120 metres. The Bab al Mandab channel is currently shallower than 120 metres, and it is likely the Red Sea was either completely severed from the Indian Ocean or nearly so during the period. This waxing and waning, severe or moderate, undoubtedly influenced the creatures then living in the sea. A time of relative isolation is usually good for speedy evolution, and there must have been such times. Even today the narrowness of the straits between Asia and Africa makes the sea relatively isolated.

After the sealing of the northern end, a couple of fish species managed to reach the Mediterranean, having presumably negotiated the complex system of waterways in use since the days of the pharaohs. Both species are able to withstand extremely salty water. Then an opening of major importance occurred in 1869, when Ferdinand de Lesseps completed the Suez canal. For the first time in some five million years Mediterranean water could mix with the Red Sea and fish could, in theory, swim between the two. In practice they could do nothing of the kind for 30 to 40 years. The aptly named Bitter Lakes, lying along the canal's route, were initially far too saline for any creature to complete the journey. Over the years the Bitter Lakes' salinity has dropped from 70 per cent to 45 per cent, a level equal to that existing in the northern portion of the Gulf of Suez. What are known as Lessepsian migrants were first seen in the Port Said area early in the current century, and the invasion has continued. Some 12 per cent of eastern Mediterranean fish are now Red Sea immigrants, such as the goatfish, the Red Sea lizardfish and the rabbitfish, all three of which are currently of major importance to Israeli fishermen.

The reverse flow, from north to south, has been nothing like so successful. Eleven Mediterranean species have been observed south of the canal, but six were still near the canal's entrance and the remaining five have not blossomed like those which travelled north. The only one ever to be caught in the southern portion of the Red Sea is the sea-bass, but even it remains much more common in the north.

The isolation of the Red Sea cannot really compare with that of the

Rift's lakes, many of which are completely severed from outside influence and colonisation. As a result, most of the fish species found within the Red Sea also exist in the Indo-Pacific ocean beyond its shores. Some 70 per cent are widespread, while another four per cent are only found in the tropical Indian Ocean and nine per cent exist nowhere else than in the western Indian Ocean. Presumably some of those which exist only near to or within the Red Sea are species which originated there and have since migrated elsewhere. The remaining 17 per cent of fish species are endemic, being found only in the Red Sea or in the neighbouring Gulf of Aden. Some of these endemics have no obvious relations living elsewhere. Since none of the early Mediterranean fauna survived, the ancestors of those fish now unique to the Red Sea have either become extinct in the Indian Ocean or are unrecognisable as relatives.

The Red Sea has its own circulation, much as happens in the oceans. Water enters via the Bab al Mandab straits. As this surface water evaporates it becomes increasingly dense and saline. By the time it reaches the sea's northern end it is sufficiently heavy to sink and, having done so, proceeds southwards. On reaching Bab al Mandab it flows over the relatively shallow bottom before entering the Indian Ocean once again, its place being taken by more water flowing in above it. This circulation system reverses when monsoon winds blow from the north-west during the summer. These encourage surface water to flow towards the Indian Ocean, and this is replaced by a deeper inflow. The stirring caused by this flowing back and forth brings nutrients either into the sea or up from its bottom. Divers may like crystal-clear water but murky, plankton-rich liquid is generally preferred by the permanent residents. The coral reefs look rich, with their diversity and colour, but are relatively starved and competition is intense. At the Red Sea's southern end, where food is more abundant, the coral reefs are much less dominant and their place is often taken by forests of sea-weeds, such as *Sargassum* and *Turbinaria*.

A further difference between the Red Sea and its neighbouring Indian Ocean is its lack of oceanic conditions. There is no significant tide, and no ocean swell to pound at the shores and reefs. There are storms but these behave differently. The Red Sea's inmates can, in consequence, be dissimilar to their ocean relatives. There are species pairs, with one species living in the sea while its counterpart, either modestly distinct or rather more so, lives within the ocean. The extent of the divergence depends on how much a particular creature needs to adapt to survive in the

154

Red Sea. If it is sensitive to the increased salinity, to the homogeneous temperature, to the lack of tide and swell, and to a different assortment of individuals, it is likely to be affected. If, on the other hand, it is like the sharks which swim, apparently casually, between one body of water and the other, no detectable differences may exist between individuals collected within the sea or the ocean. For such creatures there is no hindrance to interbreeding, and therefore no divergence.

Although the great lakes of the Rift are often rich in species, notably those of the Western Rift, the Red Sea is much more abundant. In one comprehensive survey a total of 86 fish families were investigated, which included 508 species. By comparison Lake Malawi, the most species rich of all the African lakes, has 245 species from only seven families. The paucity of Malawi families suggests a high degree of divergence to reach that species total of 245, and also shows how few species were present when the lake started upon its isolated career. Although twice as many families are represented in Lake Tanganyika, the numbers are still modest compared with the Red Sea population. Complete isolation promotes divergence within a family, with Malawi having an average of 35 species in each. Partial isolation, as with the Red Sea, permits both speciation and a steady invasion of new forms from different families, giving the Red Sea vastly more families but with only six species on average in each. A further comparison can be made with those African lakes such as Lake Turkana, which have only been isolated for a short time. Turkana has 37 species, and a mere 6 of these are endemic. Perhaps, had one been able to examine Malawi and Tanganyika a couple of million years ago, they would have had a similar total. Similarly, had one been able to examine the Red Sea when its Mediterranean outlet was closed but its Indian entrance only recently opened, it too might have contained only 37 species. Either time or immigration is necessary to build up species numbers.

Such tides as do exist within the Red Sea occur either in the most northerly section, particularly by the two small gulfs, or nearest to the Indian Ocean. They are scarcely noticeable to humans used to the Atlantic's twice-daily differences, but there is sufficient rise and fall for a modest inter-tidal zone to exist. This is populated by the standard assortment of crabs, chitons, barnacles, limpets and other forms of crustacean and mollusc. As the zone washed over twice a day is so small, relative to that existing in the Indian Ocean, many of that neighbour's inter-tidal species do not exist in the Red Sea. Moreover, the heat and aridity to which the inter-tidal species are exposed when the waters have receded, is so extreme that the Red Sea's tidal fauna and flora have to be particularly hardy to survive. Rather more devastating than the tidal shift, which does at least bring back cool and protective water every few hours, is the annual shift in level. The evaporation of summer is severe, as are the north-westerly winds, and together they cause a metre drop in the general water level. For individuals or corals which have colonised the shallows this six-month drought can be lethal. It also follows that humans wishing to see an upper section of reef exposed above the water line should choose the period of low spring tide shortly after midsummer when all the effects have combined to lower the sea.

Ras Muhammad, at the southernmost tip of Sinai, is a favoured spot for diving. It has sandy areas, some interesting wrecks, lots of coral, and a most impressive edge to the reef with a drop of many hundreds of metres, notably on the coral's eastern side. At such a diving centre it is possible to believe that all the Red Sea's underwater coast is either coral, sand, or incredible depth. However, there are also the seaweeds, already mentioned for their profusion at the sea's southern end. These can look sparse when, like some well-kept lawn, they are forever trimmed by the creatures feeding from them. To enclose a portion of rock, thereby keeping appetites at bay, is to be rewarded within a few days by a thick algal growth. This fencing is created naturally in areas where animals, such as damsel-fish, vigorously defend their territory to keep the grazing for themselves. Sea-grasses grow in shallow water, never more than 20 metres below the surface. Both the numbers of species and the quantity of grasses increase towards the Red Sea's southern end. Even so they are not as extensive as beyond the Sea. The competition for existence is so severe that there are not only many creatures feeding from sea-grass – invertebrates, fish, turtles, dugongs – but also some mimicking the grass, such as the fish *Aeoliscus strigotus*.

The Red Sea also provides the most northerly site for mangroves. *Avicennia marina* is the major species and occurs predominantly in the south, but some also exists up by Sinai. Snails, fiddler crabs, and mud-skippers are some of the species living within the mangroves. For those people who consider that the Red Sea is universally flanked by desert, the notion of mangrove swamps, with skippers and crabs exulting in the mud, is difficult to accept. Water abruptly becoming dry land is the more normal concept, and is certainly the most frequent occurrence. Sometimes the change is particularly severe, with the neighbouring shore seeming to be as harsh, arid and unwelcoming as can be found in places far removed from any sea.

Nevertheless such land can be alive with animals, simply because of the sea's proximity. Ospreys around the world will nest on almost anything – notice-boards, light-fittings, observation points – indeed anything that provides the means for achieving a little height. Around the Red Sea, notably on barren islands up near Sinai, there are no trees, no signs, and nothing higher than the glaring, shell-encrusted ground. Therefore, making an extraordinary sight, the ospreys construct their entanglements of twigs upon the land. The fishing is good, and the birds are prepared to pay this price. Occasionally, in an even more bizarre departure from normal practice, two males and one female will perform as a single pair. The plentiful food supply cannot be the cause, as a scarcity would more probably lead to two males feeding one female. One blessing in this area of general lack is that no ground-based predators exist to take advantage of the ground-based nests.

Nesting seasons are more scattered than is generally the case with birds. Some species choose high summer, when the food supply is at its peak. Others, such as the Caspian tern, prefer the relative cool of March. Plainly there is a choice between difficult alternatives: to nest when food is relatively scarce and solar radiation is less intense, or when abundant food is found in searing heat. With the summer-breeding gulls both sexes help to incubate, and there is rapid exchange when they swop roles. The chicks are also speedy in leaving their oven of an incubator, and may hurry to the cooling sea within three days of hatching.

I once landed at Assab in Ethiopia, when that country was experiencing a most difficult bout of famine. The plane was a Hercules, operated by the Royal Air Force, and was collecting 20 tonnes of food for transport to the relief centres just south of Tegre province. The sacks of flour had been brought by ship and offloaded at Assab port on the Red Sea. The harbour lies immediately to the north-west of the Bab al Mandab straits, and any plane preparing to touch down on Assab's dusty airstrip provides a first-class view of the sudden spread of water. Of course any ocean is different from the land, but the country over which we had flown, the desert wilderness of Welo with its scorched earth of dunes and rock, is so very starved of water. We had seen where rivers had grown tired of running over sand, and had died of exhaustion. We had lost count of deserted villages, where life had plainly moved from harsh to quite impossible. We had listened with more than usual interest to the engines, wondering about a landing in all that desiccation.

And then came the Red Sea, unexpected and wonderful. There was no improvement in the total drought as it approached. There was no form of hinterland and nothing like a beach to herald the immediate change. It seemed so fitting for that old Exodus story, with the water becoming dry land overnight. The Red Sea does look almost temporary, as if it was desert yesterday and might soon be again. Certainly the migrating birds consider it an aberration. If the itinerant raptors are truly loath to make a watery crossing, they really ought to cross by Sinai and Suez. They would then have dry land beneath their feet virtually all the way. Instead they persist with the old route across Bab al Mandab as if remembering that it too used to be solid earth.

·VII·
ANIMAL ENCHANTMENT
—

'You wait,' I said, 'you will see such animal sights as you have never seen before.' We had been exploring the crater highlands around the great caldera of Ngorongoro, and were descending by the road leading to the west. I became increasingly excited as, with a steady drop in altitude, the landscape grew more and more like the plains it would become. The lushness of the vegetation was gradually diminishing as we dropped down. *Acacia lahai*, the tall forest tree, became *Acacia tortilis*, the lower and more prominent form of this most African family. Rarer, but always most beautiful, was *Acacia xanthophloeia*, visible wherever slightly more water existed. As we descended towards the plains, the land became dustier and more arid, while the moist chill of the crater highland morning became the dry heat of lower altitudes. I exulted out loud at every opportunity, sometimes with no more provocation than the sight of rounded granite rocks, or dry red earth, or the sudden appearance of a giraffe so blatantly visible that one wondered why it had only just been seen. 'Look,' I said; 'isn't it perfection? And doesn't the air smell good, and feel good? As for the animals, we'll soon be seeing millions.'

We were heading for the Serengeti National Park, host to a greater number of large animals than exists in any other comparable place on Earth. It is famous, justifiably so, and is managing to survive in the modern world. The park is spread over 13,000 square kilometres, and is therefore the size of Northern Ireland or Connecticut. Unlike those two places where human beings dominate, animals rule the Serengeti and in mind-numbing numbers. Who can conceive of over a million wildebeest? What sort of area do they cover, and how can they possibly find sufficient to eat, along with all the gazelles, zebras, eland, hartebeest, and tens of thousands others that are partners to the herds? The two of us who had journeyed to see this spectacle had a vehicle, and plenty of time, and were therefore privileged beyond measure.

When we entered the park we saw all about us miles and miles of savanna. We drove, and we drove, and we saw nothing bigger than a monitor lizard. Even that scurried away speedily beneath some rocks, as if forgetful of a general instruction that no animal should be visible to us that day. We scanned the horizon, for the dust put up by many hooves, for any outline that was not flat, but to no avail. Aware of how feeble an argument it must have sounded, I explained to my companion that the total absence of herds was as much a feature of the Serengeti as the millions of its animals. To survive, the herds have to migrate, not from one place to another, but in a ragged form of the figure eight every single year. It is difficult for humans in temperate areas to know the seasons in the tropics. At home they know them well enough by looking at the trees or by sniffing at the air, but this basic awareness becomes confused when the sun shines high every single day. I had forgotten — but soon realised the error. It was August, and the great herds were then all at the Serengeti's other end. Shamefacedly, I drove to Seronera in the park's centre without having seen anything of consequence, save for blue sky, puffy cumulus, dry grass, and considerable horizon.

The following days more than compensated, and it is difficult to know which sights were paramount. Like children who can be most impressed by tea-table sparrows at the zoo, we were delighted by minor forms of excellence, such as a single serval on the hunt, or a mongoose feasting on an egg after expelling it backwards at the shell-breaking base of some convenient tree. Any recently born animal is enchanting, and we watched a baby gazelle, as motionless as death within the grass, refusing to accept that we could now see it as plainly as all along it had seen us. We saw the merest pin-pricks in the sky being fashioned into vultures coasting down, from heaven knows how high or far, to join a raucous throng upon the ground. We encountered a sick animal, walking awkwardly, and knew it could not last the night. A lion, beset with flies, growled

160

The Serengeti plains in northern Tanzania (above) are home to the greatest concentrations of large animals found anywhere on Earth including over a million wildebeest.

*The annual wildebeest migration (above left and overleaf) is one of
the Earth's great spectacles, and many predators such as hunting dogs (below
left and above), take advantage of the considerable bounty it provides.
Although the predators take their toll of the migrating herds, other
natural hazards, such as the Mara river, can claim more victims.*

Cheetahs (left) may be the fastest land animals on Earth, but not every attack is successful. Thomson's gazelles (bottom) accompany wildebeest on their migrations, and their young (below) are particularly susceptibleto predators.

The Rift's grasslands, formed with volcanic ash, have proved ideal for a wide range of mammals, such as Burchell's zebra (above right) and lions (centre and below right). The male rhinoceros lying in the foreground (above) is courting the more distant female.

The Rift Valley provides habitats for some spectacularly large birds,
such as the lappet-faced vulture (above left) and the martial eagle (below),
both the biggest of their kind. The kori bustard (below left) looks even
larger when the male displays its vivid white tail.

The highlands on either side of the Rift tend to receive most of the rain,
creating waterfalls as in the Aberdares (below left) and encouraging growth
of forest trees good for colobus monkeys (top left and above).

*A day in the Rift region of eastern Africa can provide the sights of
elephants digging for salt (below), a group of Sykes' monkeys (right),
a solitary bushbuck (far right), or even a leopard together with a
guinea fowl (below right).*

174

impotently at all the aggravation, and settled angrily to go to sleep once more.

However it is the huge herds that can be most astonishing. They thrive because of the variety of grasses, and the grasses thrive because the place is not overwhelmed with trees. In time, grassland can develop into scrubland, which gradually gives way to forest, but only, of course, in areas where trees can survive. The Serengeti plains are largely ash, blown over from the volcanic highlands to the east. Some of that ash has formed a tuff, with its grains compactly bound together, and this concretion impedes tree roots seeking deeper purchase. Moreover the ash is so porous at the surface that its high calcium content is leached to the lower levels where it forms a layer of calcium carbonate. This too obstructs the probing roots of trees. These barriers to tree growth allow the grasses to flourish for great distances. As the volcanoes are partner to the Rift, and the ash is a product of the volcanoes, and the grass is a consequence of all that ash, the tremendous herds are therefore a splendid result of the rifting process.

The scale of these migrating herds is difficult to imagine. At the last count there were over a million wildebeest in Serengeti. A large football stadium, such as the largest in Britain, can accommodate about 100,000 people. Wildebeest are substantially larger than people, and they are accompanied on their migration by several hundred thousand zebra. The Thomson's and Grant's gazelles, which either partner or follow the major herds, are also in their hundreds of thousands. It is a tremendous population on the move, making one think the grass will vanish from the wear and tear of quite so many feet. It would certainly not survive if the feet belonged to cattle, which employ less finesse in their manner of eating. The animals of the Serengeti are more sensitive and respectful. The zebras are best at exploiting the old grass material, and there is an abundance of kinds for them to choose from. The wildebeest favour the new growth, while the gazelles crop whatever is left, being able to nibble the shortest shoots. The Grants are particularly adept at such parsimony,

The crowned eagle (above left) *feeds on monkeys and small antelopes, but the slightly smaller black and white-casqued hornbill* (below left) *is a fruit eater. Both are forest dwellers.*

often remaining when the herds have gone, and finding sustenance from apparently well-worn billiard tables. If the grass is virtually non-existent they will also eat from bushes, an extra of which Thomson's gazelles do not avail themselves.

The migration occurs in spurts throughout the year but, whether on the move or relatively stationary, the herds are magnificent. If the vast army is moving, with wildebeest, zebras and gazelles forming the main body, the sight is formidable, in its sense of purpose, its numbers, and its determination. When temporarily halted at convenient grazing, the herds are spread out more widely and the whole land, from one skyline to the other, seems covered with animals. The best spots for human observers are sometimes the places which present most difficulty for the herds. They seem to know, ahead of time, where danger lies, but proceed relentlessly across the rivers, through narrower sections, over mud, and drought-ridden areas. Usually in February the wildebeest drop their young, and these infants have to accompany their walking, running, cavorting elders from almost the moment they are born. All along the route they suffer from predators which have been longing for this bounty to come their way.

For the predators the huge crowd of prey is an astonishing bonanza. They can eat and eat, and do so steadily. Unfortunately the bounty eventually moves on. The predators must then eke out a living from the small numbers of non-migratory prey. As surfeit changes to deficit, hunger becomes starvation, particularly for the very young and the old. This is the prime reason why predator populations in the Serengeti never build up sufficiently to make serious inroads into the tremendous quantity of prey. The carnivores undoubtedly affect the herd's numbers, but the physical obstacles, such as the crossing of the River Mara, can create a greater slaughter, keeping crocodiles full-bellied for many weeks. The migration is a year-long obstacle course, lethal to many of its participants. For the herd as a whole it is salvation, and that matters most. The wildebeest is not an obvious survivor, being neither very fast nor very strong, but its numbers are formidable. The herd's way of life is undoubtedly well attuned to this grassy, dusty, beautiful area lying downwind of the volcanoes that gave it birth.

A quarter of a century ago I had the good fortune to live under canvas in this most magnificent park and, at the conclusion, I wrote:

Having seen unforgettable things the sole desire was that anything and everything should be done to conserve the Serengeti. Perhaps it had been 30,000 wildebeest thundering past a spot that had sparked it off, or a pack of hunting dogs, or a baby zebra with the light and the world behind it, or a solitary topi standing sentinel on a mound; wherever we went in the Serengeti, and whatever happened, there were sights to be seen that were immemorial and wonderful. We came back each evening heady with enchantment and longing for more. Like anyone who goes there, we were immediate converts to its cause. The Serengeti is a legacy that must always be. Whatever the difficulties, it must survive: its destruction is unthinkable. For anyone who imagines otherwise, let him go there, and let him be enriched by it.

Fortunately, in the quarter century since that time, the Serengeti has survived. When colonial rule of the area was drawing to a close, many expressed forebodings about the fate of the great herds, assuming that the animals' days were numbered. Bernhard Grzimek and his son first exploited an aircraft to count the wildebeest in the late 1950s and reached a Serengeti total of 99,000. A further aerial survey three years later boosted that figure to 210,000. The current population is over a million and it has not dropped below that quantity for several years.

Of course there have been setbacks – politics, poaching and population being three critical headings – but the animal reserves linked to the Rift are still intact. I was delighted to see six rhino and a small group of elephants in Ngorongoro on my last visit in 1987, but such creatures are now a rarity where once they were a happy commonplace. A recent aerial survey of the Galana Ranch area in Kenya counted 90 live elephants and 425 recently dead, compared with a 1972 total of 4379 living animals in a similar series of flights. There were 22,174 elephants within the Kenyan Tsavo park in 1973 and the current figure is 4327. All big animals living beyond the jurisdiction of game reserves are less numerous than they used to be.

Despite all this, the visitors who arrive up and down eastern Africa to see the sights leave gratefully, and give thanks they were so privileged. Some even remember that their sense of debt should be levelled at the Rift. Without its contortions, its slips, scarps, and faults, and without its volcanoes with all their lava and ash, there would not be such a profusion and confusion of different kinds of animal. The fracturing of the land has created abundant habitats, and life in all its forms has responded to the full.

The Rift Valley has not only been responsible for enormous concentrations of wildlife; it has also acted as a barrier. It is far from being an insurmountable obstacle, but its combination of escarpment, lake, change of altitude and shift in climate has been sufficient to cause some species to develop distinct western and eastern populations. The wildebeest, for example, could certainly cross such hindrances, for they can swim, scramble down slopes and are used to desert dust beneath their hooves. However, the individuals on the eastern side of the valley are conspicuously bigger than their western counterparts. Grant's gazelle, another seemingly ubiquitous animal in eastern Africa, is also affected by the Rift, and a different sub-species exists on either side of it. With waterbuck there are two separate species, Common and Defassa. The more easterly Common Waterbuck possesses a double white patch on its buttocks. *Kobus defassa* has an elliptical white ring instead, and exists in western areas, notably Uganda. The enchanting, tail-wagging Thomson's gazelles are fairly uniform across both sides, east and west, but individuals do not appear to cross the Rift.

Perhaps it is the general suddenness that can be so disarming. Most habitats merge gently into one another, without a savage contrast. In much of the Rift there is an abrupt alteration of its many variables. As the rainfall lands mainly on the higher altitudes, and therefore causes more vegetation, the drop of an escarpment can lead to increased temperature, much less growth, an arid landscape, less concealment from predators, and a less appealing situation. In theory, temperature shifts by only 1° Centigrade with every 200 metre change in altitude, but the Rift seems to make nonsense of that general truth. At its base it can resemble a frying pan in its heat and airlessness. Less than 1000 metres up lies another world, refreshing in its breeze, rich with shade, and with an appeal that must affect animals as much as people.

A further peculiarity of the Rift is that its habitats change over time. A lake may come or go. A fresh outpouring of lava can initially devastate and then create a different kind of world. The sudden droughts and downpours, such a feature of eastern Africa, have their effects highlighted by the Rift, as this arid zone becomes a desert, or this area receives just sufficient water for trees to survive and attract more rain. Even without the tinkering of human beings a series of photographs taken at quarter-century intervals would show dramatic changes, in the vegetation, in the animals, in the habitat. Many portions of the world would show no detectable differences as the centuries merge, each one into another, but not so the Rift.

Some of the creatures seem to expect such changes, and behave accordingly. Friends of mine were excavating tree-roots about two metres below the surface in the harsh terrain near Lake Turkana when they encountered resting, barely metabolising, mucus-covered frogs. Any downpour will cause such portions of the Rift to become awash with amphibia. The effects of the rains are short-lived and, when the puddle or lake vanishes as speedily as it came, the frogs burrow in the mud and there cocoon themselves. Ten years may elapse before another bout of water awakens them from their dormant state. Then any humans, who may have been kicking the dust in that desiccated spot, are abruptly surrounded by hundreds of frogs which appear to have descended with the rain.

The Rift was created sharply and dramatically, and the landscape is as magnificent as the animals which dwell in it. The quails leap seemingly for ever, as one's wheels hum over the Serengeti sward, but they do so against the backdrop of the crater highlands. From these higher altitudes, as buffalo or elephant lumber across the track, one can see down to the plains as clearly as from the plains up to the hills. There is always supplement: a volcano relieves a flat horizon; a lake glistens within a desert; a kopje bursts suddenly from a great expanse of green. The animals and plants have taken every opportunity the varied Rift has afforded them. A piece of basalt, stark and upright, is home for vultures and lammergeiers. A soda lake, unsuitable for fish, is supreme for flamingoes. Even a hot spring, as at the southern end of Lake Magadi, is the favoured habitat of a small fish. *Tilapia grahami* can exist in water up to 41° Centigrade, about the temperature of a good bath or a terrible fever. One wonders if any portion of the Rift system exists that is unexploited by some cunning plant or animal.

Because the Rift is a sinking down, a lowering, it is easy to forget the higher land on either side of the sunken portion. These shoulders are sometimes taller than they were before the Rift was formed, forced up by all the lowering. For example, much of the Aberdare mountain range in Kenya's central highlands is above 3000 metres and some peaks are higher, such as Kinangop at 3906 metres and Ol Doinyo Lesatima at 3999 metres. Far from being hot and arid, the Aberdares are so replete with water that, in various favoured places, it cascades downwards, hundreds of metres at a time, in magnificent waterfalls. The mountainous shortage is not a lack of water but of salt, with areas rich in this equally precious commodity being centres of attraction. To drive up past the bamboo, to put on another sweater or two, and to emerge on a plateau that any Scotsman might think of as his own, is to be splendidly bewildered. If this is Old Caledonia, what is that elephant doing? If it is cold and damp (which it surely is, particularly for over-nighters) why are there buffalo standing by that pool, seemingly unmindful of a drop in temperature and rise in altitude? To enter the forests is possibly to see the giant forest hog, a forest animal unexpected in the upper reaches of a mountain.

To add to the confusion, many of the animals do not even look like themselves. Servals, frequently seen on the open grassland of the Aberdares, are often so black that they are far more visible than the normal varieties of their kind on the plains below. Augur buzzards seen at altitude can also be black, having lost their traditional white front. And the leopards, as happens in other parts of the world, lose the blatancy of their spots as they too adopt the melanistic form. Old male bushbuck and various forms of genet can be nearly black up in the Aberdares, and it is easy to wonder at such widespread darkening. It certainly does not assist in camouflage, and is a presumed response either to increased solar radiation or to cold. With a part of Africa looking like Scotland, and with many of the animals not even looking like themselves, the visitor should remember there is nothing commonplace about the Rift. As the books frequently say: 'Its habitats are diverse.' They are indeed.

The Ethiopian highlands are even stranger, in their environments and inhabitants, than the Aberdares. Ethiopia itself is a remarkable portion of Africa and the Rift. The country covers only 3.4 per cent of Africa's total land area, but it contains 50 per cent of African land above 2000 metres and no less than 79 per cent of the land above 3000 metres. There is higher ground elsewhere in Africa, but nowhere else are mountains quite so extensive or so varied in their climate. It might therefore be expected that Ethiopia's massif contains animals unique to the area, and such is certainly the case. Of the 214 species of mammals in the country as a whole, 23 are endemic, the majority of which live at the higher altitudes. Even the bird list for Ethiopia contains 24 endemics out of the breeding total of 665 species.

The most intriguing animals of the Ethiopian highlands can be seen in a park lying 400 kilometres to the south-east of Addis Ababa. It is in the area of the Bale mountains, and contains moist tropical forest, high peaks (such as Tullu Deemtu at 4380 metres) and a quantity of high altitude forest and grassland. Africa's highest all-weather road runs through the park, from north-east to south, and is the only such highway on the continent to traverse the 4000 metre contour.

The bird life is complex enough to muddle any ornithologist taken blindfold to the area and then exposed to its various offerings. There are European raptors over-wintering, such as the kestrel, marsh harrier and pallid harrier. There are locals, like Verreaux's eagle, the tawny eagle, the augur buzzard, and the long-eared owl. Various passage migrants, including the white stork, black stork and European bee-eater, can add to the confusion. The wattled crane breeds here, and a rail that is only found in Bale. As for the mammals, the area possesses the mountain nyala, an endemic distinct from the nyala of south-eastern Africa. There are endemic sub-species of bushbuck (known as Menelik's) and some eight of the park's mouse-like and rat-like rodents are endemic. However, perhaps the greatest oddity is *Tachyoryctes macrocephalus*, the giant mole rat.

There are 14 species of mole rat in eastern Africa, but the giant form is unique to Ethiopia and is confined to its mountains. This large rodent weighs about a kilogram and lives almost entirely underground. It was first described by a German, Rüppell, in 1842, but was hardly seen for more than a century afterwards. In the late 1960s it was again observed 'in the field', as the saying goes, but with more reason in this case as the animals turn their locality into a passable imitation of a badly ploughed field. Unlike a true mole, which feeds below the surface, the mole rat tries to achieve the best of two worlds by feeding from the richer growth on the surface while maintaining subterranean security. It tends to keep its back half within the hole and uses its large, powerful incisors to tear at any grass within reach. Only very rarely does this animal leave its hole, as if mindful of the many predators waiting to swoop from the heavens. The food is not eaten on the surface, but pulled below and consumed at leisure out of sight of enemies. The hole also provides warmth, which is necessary because the surface at Bale freezes every night. The animals wait until about 3 hours after dawn, when the frost has disappeared, before beginning upon their cautious, front-half-only, foraging.

Each mole rat uses more than one hole, presumably because the food supply within reach of any single outlet is so limited. All this burrowing makes the existence of the giant mole rats immediately detectable. The evidence is supplemented by piles of hay, for much of the vegetation dragged below the surface is rejected at a later date. Adding to the impression of a dismal attempt at agriculture is the fact that once the species has encountered a suitable piece of land it makes fulsome use of it. One biologist has estimated that there are 6250 mole rats living in each square kilometre. As there are also many other forms of rodent, some endemic, others not, it is understandable that such a zone is rich with predators. For birds of prey there is considerable food down there, scurrying between the piles of hay, and grubbing at available grasses while ready to retract at any hint of danger. The mole rat's eyes are towards the top of its head, and its hearing is extremely sensitive. Since the animal rarely strays far from its burrows, it seems immune from predation. However, one important carnivore knows well how to catch mole rats – it merely bides its time.

The Simien fox is endemic to Ethiopia, and is one of the Bale park's most famous inhabitants. In fact, this handsome foxy red animal is not a fox at all but a long-legged, long-snouted jackal. Bigger than hunting dogs or other jackals, it only lives in mountainous areas. The Simien fox's hunting technique is to take short steps, sometimes in a crouched position, and then freeze, stalk and freeze again, until eventually it pounces. It is a past master at patience, and has been observed to freeze for 10 minutes at a time. Sometimes, it will dig out a burrow, but surface catches are more common. Very occasionally it chases prey, such as cape hares, but most ineffectually. Although the surface temperature often drops at night to minus 6.5° Centigrade, these fox-jackals do not make burrows, but sleep in the freezing open. Almost nothing is known about their reproductive cycle, such as when the pups are born and how many there are per litter, and Simien foxes seen in the daytime tend unhelpfully to be single individuals.

The same Rüppell who described the giant mole rat also identified the Simien fox in 1835. Since then it has been frequently seen and frequently killed. There are unfounded suspicions that the animals kill sheep and that their livers have medicinal properties, but agriculture remains the greatest threat as human farming can take away the Simien fox's livelihood. It is unlikely to do so among the plateaus of the Bale mountains, owing to the high altitude and night-time cold. Current numbers are thought to be about 750, with the majority living, and eating rodents, in the Bale park. So this unique Ethiopian mammal, driven from so much of its former range, may have found a final sanctuary among the largest of lobelias and the biggest of mole rats.

In my experience no one visits the Rift areas of Africa, intent on seeing the wildlife they have to offer, and returns home disappointed. It may be the numbers that have impressed the most, or some single spectacle, or the strangeness of creatures found nowhere else on Earth. Whatever the sights, multitudinous, solitary or novel, there is always a portion of the Rift as backdrop to the scene. Without it there would be a uniformity which would quickly pall. With it there is a seemingly inexhaustible supply of astonishments.

· VIII ·
APEKIND TO HUMANKIND

There is one important creature that should be added to all those mentioned in the previous chapters. The Rift may have been the backcloth to its evolution and also the spur, permitting and encouraging – indeed, requiring – a superb ability to adapt to a variety of environments. For a modern visitor journeying up and down this region it is difficult not to be aware of the extremes of habitat that it contains – and the existence throughout of this very successful animal, both now and in many millennia past. It is fascinating to think that the Rift was the setting for the rise of *Australopithecus* and *Homo*.

These two kinds of erect ape, together known as the hominids, apparently conspired with the Rift's geology to write a detailed history for descendants to read, as many managed to die in some of the best conditions for fossilisation anywhere in the world. The slow sedimentation of eastern Africa's rivers, combined with the changes in drainage and faulting that accompanied the tectonic shifts, has created a palaeontologists' bonanza. As if that were not enough, the best of the fossil beds are unencumbered with vegetation and are constantly renewed, as if pages are being turned, by erosion exposing new layers.

It is appropriate that the Rift's fluid geology has recorded the history, for it was the geology that created the conditions for hominids to evolve in the first place. Four million years ago apes were already characterised by the fact that their forelegs had developed into specialised appendages,

186

with uses far beyond simple locomotion. The forefeet could grab things, hold things, pick things, throw things; but they were still used for their original evolutionary purposes, such as swinging through trees or the kind of knuckle-walking favoured by gorillas and chimpanzees. *Australopithecus* was an ape that used its forelegs exclusively for other things: it had true arms and hands, and progressed only on its hind legs.

Why? There are many theories, but most relate to the fact that the hominids were travellers. The continual stretching of the landscape gradually turned what had been primarily forest, still suitable today for chimpanzees and gorillas, into a region that was primarily plains, with sparse woodlands, isolated forest patches and lakes. Every animal that is native to the Rift Valley is the result of a species successfully exploiting this new geography – the great grazing herds, for example, or the monkeys that remained in the forest patches and continually subdivided into new species, each specialising in each patch as though it lived on an island. And the hominids made a living that involved covering the distances between trees, rivers and lakes. For that they needed to move comfortably on the level, and, in terms of energy, cheaply. They could use their arms to carry things, they were adaptable, and they walked.

The earliest known hominid, *A. afarensis*, barely distinguishable from a modern chimpanzee except for its upright posture, left remains in Ethiopia's northern Afar depression, in southern Ethiopia at Omo and in Laetoli, Tanzania. Evidence of later species, *A. africanus* and *A. robustus*, has been found in southern Africa, outside the valley altogether. At about the same time, 2.5 to 1.5 million years ago, a second upright-ape genus appears – *Homo*. Its first known species, *H. habilis*, has left the earliest evidence of itself, mainly in the form of chipped-stone tools, at Koobi Fora by Lake Turkana, at Olduvai Gorge near the Serengeti, and, again, at sites in southern Africa – the first and last being more than 3000 kilometres apart. It would be a coincidence of astronomical odds if those places also represented the extremes of *Homo habilis's* range.

The age of *H. habilis* finally gave way to that of the large and brainier *H. erectus*, predominant from about 1.6 million years ago until about 200,000, when it began to be replaced by *H. sapiens*. Both *erectus* and *sapiens*, in their respective times, branched out of Africa altogether, but that was expansion, not desertion, as both species also stayed behind. This does not mean that the *H. sapiens* who live in the Rift Valley today are necessarily descendants of the ultimate human stay-at-homer – many of their ancestries can be traced to other places in Africa – but they are humans living in an extremely varied valley most suited for adaptability. What is striking to the modern traveller in the Rift is how many kinds of habitats there are and how easily humans can bend to them. In a few kilometres it is possible to meet distinct communities, each having learned to make the best of a distinct area. They may be pastoralists or hunter-gatherers or cultivators or fishermen. They may live in the light and the heat of the plains or the dark and the relative cold of the forest depths. They may be highlanders or lowlanders. There are desert people and lake people, nomads and villagers, farmers and hunter-gatherers.

It is often thought that these last have the most primitive life-style. It is certainly true that such activities as herding and farming are relatively recent ways of using nature, and other ways, such as fishing, can be constantly updated. That bushmen seem to be behaving as they have for centuries does not have to mean that the very earliest *Homo sapiens*, which date back *two thousand* centuries, behaved that way. But, on a fundamental level, hunter-gatherers seem to be more a part and parcel of nature than some of the other people in the region, and it is hard not to be curious about them.

Once I even made an attempt to find some, in particular a group known as the Hadza, a small and cautious people who live by Lake Eyasi. Having no particular reason to want to be found by me, they kept firmly out of sight. On retracing my steps, I saw their neater footprints superimposed on mine, and that was all I saw of them. As a result, I was even more enchanted when I first heard, in the late seventies, of the discovery of the Laetoli footprints. These had not been left by *Homo sapiens*, or even by *Homo*, and their scientific significance lay in the fact that, being 3.6 million years old, they provided the earliest evidence of bipedalism. They had been made by the original hominid, *Australopithecus afarensis*, the walking chimpanzee, and were left by three individuals who, one day in the life of the planet, had been walking along together. One was a child next to an adult while the other adult followed behind, obscuring and duplicating the first adult's footprints. They walked on ash, and a further scattering covered the marks until modern *Homo sapiens palaeontologicus* arrived to brush the ash away. Beyond the footprints lie the old volcanoes, the probable source of the ash, and it is splendid to be confused between then and now.

Even though it is important to emphasise the differences – in species, in genus, in social behaviour, in the vastness of time – between the earliest hominids and the present one, there is another perspective that should not be ignored. If we take a grouping that includes the chimp *Pan troglodytes*, the pigmy chimp *P. paniscus*, the gorilla *Gorilla gorilla* and all the hominids through *H. erectus* and the subspecies of *H. sapiens* that we call Neanderthal, and throw in *H. sapiens* itself, we have a range of genetic difference that is not as great from one end to the other, from gorilla to human, as between a polar bear and a brown bear. There are some biologists who argue that if we were more honest with ourselves, and examined ourselves and our relatives in the same cold light that we use for the other animals, all of the African apes, including us, would be in the same genus. This knowledge comes from the relatively recent techniques of molecular biology. These techniques are now being used to assess the evolutionary relationships of modern peoples throughout the world, and from them two things have been learnt. One is that all humans are very closely related to each other, with a recent origin for modern humans; and the other is that their origin lay in Africa (about 100,000 years ago), possibly even the Rift.

I was never able to find the Hadza, but two friends and I once had the opportunity to visit some of the other current inhabitants of what is now believed to be humanity's true motherland. These were the pastoralist Masai, whose hospitality we were delighted to accept one evening after our vehicle had failed. We were desperately hungry and thirsty, and were given mugs of milk – among the most refreshing drinks I have ever tasted, and, half-crawling, half-stooping, were led through the doorway of a dung dwelling. We were amazed at the cosiness inside. There were a couple of women together with their babies, and a small fire was burning at the centre. A cow's crackly hide had been spread for our convenience, and we settled to pass the night. Outside were all the cattle and goats. Beyond them was the circular enclosure of thorn and bits of tree to keep predators at bay. We had never been in such a place before, but were comfortable and at home.

The next day I went to the nearby *Homo erectus* hand-axe site of Olorgesailie. There are many hundreds of these implements lying around and they are called hand-axes as a kind of shorthand: no one knows their purpose. All are more or less pointed at one end, and more or less rounded at the other. Why these tools should be found in great repositories such as this is not known – perhaps it was a factory, perhaps a rubbish heap.

Some palaeontologists have suggested that what we assume to be tools are, in fact, the leftovers, and that the chips were the real tools; but whatever the significance of the site, it had not been used for some 700,000 years. Nevertheless, when watching an archaeologist digging in the dust, it was easy to imagine that he was making the implements rather than merely uncovering them. The countryside looks so right for an earlier existence. There is an occasional giraffe, small thorn bushes on every side, rock, earth and not much more. In *Homo erectus*'s day a vast sheet of water filled the lower contours of this land. This has since disappeared – just as so many Rift lakes have come and gone – but in every other way the scene is much the same, with people hacking at the dust, with hand implements everywhere and a great feeling of continuity between then and now.

This sensation is greater up at Koobi Fora on the north-eastern shore of Lake Turkana. The ebbing and flowing of the lake, together with most cooperative erosion, has helped to create and expose a superfluity of fossils. Expert eyes are required to locate a hominid fragment, which may be no more than the hint of a cranium bulging through the soil, but anyone can spot the abundance of other fossils – perhaps a mussel or a piece of vertebra, a crocodile's tooth still with its shiny enamel or a piece of leg-bone from an antelope. Sometimes it is hard to walk without treading on some part of an earlier living thing, and the ages of the fragments can vary from a few million to hundreds of thousands or a mere handful of years.

An equally venerated spot is Olduvai Gorge. Named by Masai after a wild sisal that grows there, this most prolific of fossil beds is some 100 metres deep in places and about 40 kilometres long. It lies on the edge of the Serengeti plains, and its abundance of fossils reflects the vast number of animals grazing and browsing on every side; but the creatures unearthed from its Pleistocene strata are as far removed from all those standing on the plains as are the small hominids from the modern people who live or work nearby.

Every creature that evolved here had to adapt to changing circumstances. There was a lake here once. Over the ages, it expanded and contracted and, for the time being, it is gone. Ash fell whenever a nearby volcano reached its bursting point, and the climate varied from wet to dry. The pigs, for example, dramatically altered their teeth as they changed from browsers to grazers. Most animals went through rapid successions of speciation. The hominids did this too, until finally a couple of models emerged with a huge enlargement of the very organ of adaptability, the brain. *Homo erectus* and its successor, the even more adjustable *H. sapiens*, proved capable of living in virtually any environment the Rift could offer. And being fit for the Rift made the rest of the planet easy pickings. First *erectus* and then, 100,000 years ago, *sapiens* burst across the land bridges of the Red Sea and expanded into plains, forests, deserts, mountains. These humans could live in cold places, far from the habitats of any other apes; some even settled in the high Arctic. They could float on the seas. In time, they could fly. A few have even managed to leave the planet altogether.

The Rift made humans so adaptable that they have become a problem. Their next test of adaptability is whether they can adjust to the fact of their own existence, whether they can solve the environmental problems that they themselves have caused – or whether they will fail at last and take the planet down with them.

·IX·
NATURAL BLESSINGS
—

There are disadvantages for the nations of the Great Rift, such as steep escarpments, difficult lava, active volcanoes, surface ash, faults, and abrupt changes in altitude. There are also advantages. The heat so plainly existing below the surface is one of them, provided it can be tapped efficiently. The substances belched forth can be profitably mined when nature has concentrated them in a convenient manner. The soil, based on its laval origins, can be extremely fertile and, provided there is water, will produce good crops. It is not all bad news when a portion of the Earth's crust is torn apart.

One particular benefit of the fractured area around Lake Magadi has been most effectively exported for three-quarters of a century. When the railway line from Mombasa to Lake Victoria was conceived in the last decade of the nineteenth century, with the prime purpose of connecting Uganda to the sea, no one thought of the traffic it might bear. Uganda had little to export, accustomed only to porters carrying head-loads to the coast, and the intervening landscape was equally non-productive. Once the lunatic line, as it was called, had been built, attitudes began to

Among those animals that can survive in the more arid areas of the Rift are the beisa oryx (above right) *and the gerenuk* (below right) *known in Swahili as the 'antelope-giraffe'.*

Few plants can survive the conditions existing in the Bale mountains of south-eastern Ethiopia. One that flourishes extremely well is the giant lobelia (below) whose spike can reach 6 metres. The appropriately named beret plant or Echinops langisetus (bottom) *is found in the Ethiopian highlands.*

194

Mountain nyala (below) are found only in Ethiopia, the last large mammal to be discovered in Africa. Black servals (bottom) are found in mountainous areas and are very rare.

Gelada baboons (above) live in mountainous areas, and can travel in bands of up to 600 individuals. The bleeding-heart baboon (inset) is the largest primate apart from the great apes, and lives mainly on grass.

The Olduvai Gorge in Tanzania (above) has been a rich source of fossils, including some of early hominids. Footprints found at Laetoli (right), identified as 3.6 million years old, are the earliest proof of bipedalism in the ancestors of humankind. The hand-axes (far right) were made some 500,000 years ago.

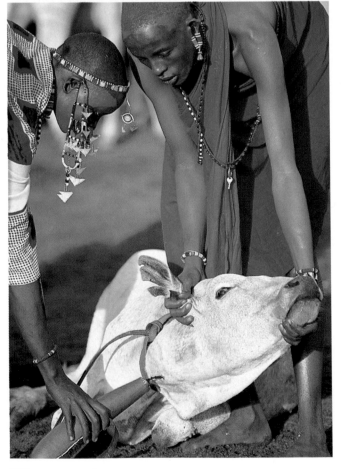

The Masai are a pastoralist people living in the Rift region who have largely kept to their ancient ways, whether adorning their faces (above left), tending livestock (above), or extracting blood from living animals (below left).

Lake Bogoria (left) *lies to one side of the magnificent Laikipia escarpment, and is fed partly by hot springs* (below) *that bubble from the rocks on its western shore.*

Hot springs (overleaf) *can be tapped to create economic electricity if the steam can be located in sufficient quantities at a reasonable depth. Until that happens what could be a valuable asset merely blows off into the atmosphere.*

There are advantages to a volcanic area. Drilling for steam (below) can lead to cheap electricity, the soda lakes can have their salts extracted (bottom) and the farming land can be extremely fertile (right).

change. It was ridiculous merely to carry ivory and other such items that had made up the previous trade. Railways are bulk carriers, and suitable bulk had to be found to satisfy the new line's demands, eventually repaying the cost of its construction. Agriculture was the obvious choice, and the East Africa Agriculture and Horticulture Society was formed in 1906.

European settlers were already farming 460 hectares, but they were growing too little of too many kinds of crop. Staples were what was required. The railway would be a blessing for such crops, and such crops would be a blessing for the railway. The lunatic line would lose its lunacy the moment that exports rumbled profitably down the track. 'I want traffic for the railway,' said the British minister for the colonies on a visit to the region in 1906. 'Something that the world wants and which will be a staple, reliable and increasing traffic.' He too thought in terms of agriculture, and settlers were encouraged to settle inland, around Machakos, Thika, Naivasha and Nakuru, rather than at the coast. The climate up-country was preferable for Europeans, and every other form of inducement was used to encourage farming inland and away from the sea. Since the railway existed it must be used, and the settlers were persuaded to grow crops in areas where their export was impossible save by train.

During the earliest years of the twentieth century no one realised that an excellent product was waiting to be exported, and not too far from the railway. Soda did not need to be grown. It created no problem of land acquisition. The world had need of it and its production would be entirely reliable, susceptible neither to climate nor the pests that could destroy many a farmer's hopes virtually overnight. Lake Magadi had been observed by Europeans since the early 1880s, and many of these visitors must have gazed upon its shining whiteness even if they did not actually bother to investigate this unwelcoming and apparently infernal region. The lake is sizeable, being 26 kilometres from north to south and about eight kilometres wide, but it lies off the main route and is only 600

Lake Magadi receives its water from soda-rich streams (left),
and evaporation causes the soda to form a crust (inset) *thick enough
for trucks to drive on. Half a million tonnes of sodium carbonate are taken
from the lake each year.*

metres above sea level. Anyone who does venture down to the shimmering lake is unlikely to stay long, with hot soda springs by the shore, no visible plant growth, and an all-pervading chemical smell.

In 1909 a small exploratory party decided to investigate Magadi. The lake lies 85 direct kilometres from Konza, the nearest point on the railway line, and is some 900 metres lower. The chemist in the party discovered Magadi to be an incredible source of soda, containing possibly in excess of 200 million tonnes. The deposit is largely sodium sesquicarbonate, sometimes called trona. When processed this can be turned into sodium carbonate, which is a raw material for many industries, such as soap and glass. To use the colonial minister's words, it was 'something that the world wants', and was entirely right for the railway. Although the main line from Mombasa to Lake Victoria had only been completed at the very end of 1901, and had not begun to pay its way, there was immediate talk of another railway leading to Magadi. This second track would cost more money, but would help to pay for the first. Work began in 1911, after the Magadi Soda Company had been formed, and the necessary 146 kilometres of line were completed in 1913. Specialised docking facilities for the new cargo were constructed at Kilindini, the port built near Mombasa on the railway's behalf. An initial estimate for the first year's production was 80,000 tonnes, the sort of figure suited to a railway, and out of all proportion to the elephant tusks and similar merchandise brought to the coast in the pre-railway days. Magadi was a salvation no one previously had envisaged.

In fact the first world war interrupted early export hopes, and they were not realised until 1920. In that year the value of exported soda exceeded both sisal and maize, equalled hides and skins, and was only beaten by coffee. The percentage value of soda in Kenya's export economy then dropped, mainly because agriculture proved to be such a rich provider, but the Magadi enterprise has been steadily lucrative. On average, a couple of trains leave that 600-metre level each day, struggle upwards to pass through Kajiado on the road to Tanzania, and finally reach the junction and main line at Konza. The trains pull 20 wagons apiece, and each wagon carries some 38 tonnes of soda. Output is about half a million tonnes a year, and 1988 saw the seventy-fifth anniversary of the first shipment. Extraordinarily, the Magadi deposit does not seem to have been diminished by the steady removal of so many million tonnes of soda.

Another lake on the Rift system, the Dead Sea, is world-famous for its buoyancy. Visitors photograph each other reading newspapers in its waters to emphasise the lake's considerable saltiness. The river Jordan flows into it and, as there is no outlet, brine has accumulated over the centuries. At Magadi there is also no outlet and visitors also photograph each other by its shores, but there the similarity ends. Magadi's accumulated chemicals are so concentrated that anyone wishing to read a newspaper can stand to do so, and trucks can be driven across much of its surface without fear of vanishing. Perhaps the word 'lake' is a misnomer. Magadi, producing soda that does not diminish and bearing the weight of trucks on its surface, is clearly no ordinary lake. In fact there is nowhere else like it in the world.

The sodium carbonate it possesses is directly linked to volcanic activity in the area and hence to the rifting process. To its south is the still active peak Mount Lengai. The effluent from this volcano is largely carbonatite which is particularly rich in sodium carbonate. Other volcanoes give forth some carbonate of soda but Lengai is the only active volcano in the world to produce carbonatite. Various East African lakes contain soda, such as Natron, Manyara, Nakuru and Bogoria, but not to the same extent as Magadi. It is therefore assumed that the rocks below both Magadi and Lengai contain much more sodium carbonate than is normal for volcanic regions. The streams flowing into Magadi are extremely rich in soda, and as several million tonnes have been extracted without denting the resource it seems that the area's quantity is prodigious. The Magadi Soda Company seems assured of a steady future.

The Rift Valley is, in general, very fertile. Initially, the Earth's effluents are extremely destructive to life but, given a few centuries and suitable erosion, the resultant volcanic material possesses most desirables for plant growth, except for nitrogen. Over much of the Rift Valley there is often a lack of vegetation, and therefore an apparent infertility, but the rains provoke a tremendous flowering. Even without such rainfall there is a great arena of colour to the south of Lake Naivasha, the cause being carnations. The reason they can thrive at all, let alone in such profusion, is because they are watered with some 7 million litres a day taken from the lake.

If anyone thinks carnations are simply pink or white, they should walk for a couple of kilometres along Sulmac's plantations, where they will pass nothing but carnations of some 50 varieties. Until 1972 all those

Sulmac hectares south of the lake yielded whistling thorn and not much else. Nowadays the passing wind cannot create that same uncanny sound. Instead there is a rustling from 120 hectares of carnations and huge hangars of plastic greenhousing. The old days may be missed, with the occasional gazelle or ostrich wandering past the thorns, but 200 million stems of plants exported every year are a profitable exchange. About 40 *tonnes* of flowers are loaded each day on to a freight-carrying DC8 that speedily carries them to Europe, principally to the Federal Republic of Germany.

The flowers undoubtedly thrive at this 1800-metre altitude, which possesses the right degrees of heat and cold. Carnations are the major crop but there are also *Alstroemeria* and other lilies, *Liatris*, *Ornithogalum*, and many kinds of rose. The Naivasha litres of water are plastic-piped below the surface, thus permitting a 30 per cent saving over aerial irrigation. Fertilisers are included in the water and the plants are also sprayed with pesticides and fungicides. This is all very satisfactory for the blooms, but creates concern about the lake. Not only is it losing water but possibly gaining unnatural chemicals from run-off during heavy rains. Tests affirm that nutrients are finding their way into the depleted waters, perhaps from other sources. The water loss is undeniable but, according to the carnation farmers, Naivasha was 1900 metres above sea level in 1890 and only 1880 metres by 1954. It rose tremendously after 1962, and the lake may only be experiencing a similar fluctuation. Its average depth is about 8.5 metres, and there must be a natural drainage to account for its freshness, with no surface streams acting as overflow. The litres which quench the carnations' thirst form an impressive total but, taking into account that water is not extracted every day, only lower the lake by half a centimetre a year. Evaporation must account for many times that quantity, perhaps a hundred times or so.

When a lake is receding, when mud replaces the normal bordering, and when the usual plant and bird life alters quite dramatically, it is inevitable that any obvious water loss is adversely criticised. Perhaps all the farms around Naivasha, including the flower farms, are taking a significant and damaging quantity when the years are added together. Maybe the uphill farmers, using the streams that feed the lake, are taking much more than in former years, or the felling of trees at still higher altitudes is having a drying effect, or perhaps the Olkaria geothermal station, just down the road from the carnation property, is blowing off some Naivasha water along with all its steam.

To an outsider, geothermy, the harnessing of the Earth's excess heat, seems an excellent idea that should present few problems. Everyone knows, particularly miners, that temperature rises with quite modest burrowings underground. Around the Rift this fact can be conspicuous, with steam actually curling from the surface. The inside of dormant Longonot has steam vents on its northern crater wall. On Mount Suswa, where surface water is in short supply, the local Masai use a simple condensation procedure to gain the precious liquid. By placing a flat sheet of iron, say a piece of petrol drum, over the vent they are rewarded, at the iron's bottom end, by a steady dripping of hot, slightly sulphurous, foul-tasting but drinkable water. The *maji moto*, or hot springs mentioned earlier, do not appear only on the western shoreline of Lake Bogoria. They also surface, for example, on the western shore of Lake Manyara, and at the southern and northern ends of Lake Magadi.

Since steam and hot water are so visibly present, they must exist in greater quantities below ground where there has been less opportunity for cooling. The creation of electricity generally demands the boiling of water to make steam, and as one watches the Rift Valley's excess heat bubble to the surface, it does appear that the task of electrical generation has already been half achieved. More importantly, the expensive part of the process is normally the costly fuels such as coal or oil used to create the heat and steam. Having nature produce the steam makes the equation look so good that one wonders why this Earth heat, so free and so available, is not a basic power source in every country, even those that have no visible steam upon the surface of their land. Unfortunately nothing is ever quite that easy. The heat is free, it is available, but harsher realities can make this gift uneconomic to accept.

There has to be what is known as a high heat flow. Drilling through the crust is expensive and it is important to reach hot rocks quickly. In Britain, for example, one experimental bore-hole proved the rise in temperature with depth, and the results were quoted by Celia Nyamweru in her most useful *Rifts and Volcanoes*. At the surface, on a cool day, the rock was at 6° Centigrade. At 400 metres, the temperature had risen to 27° Centigrade. This is a sizeable increase, but still no more than a warm day in many parts of the world. At double this depth, the temperature had risen to 40° Centigrade, a further leap but still no hotter than a feverish body. To achieve 150° Centigrade – a heat worthy of extraction – it would be necessary to drill 6 kilometres down, and much of the heat would then be lost on its passage to the surface. There would also be the problem, and expense, of sending some medium to gather this sub-terranean heat, and probably water would be pumped down to be transformed into steam. Suddenly it is little wonder that people in Britain mine coal available much nearer the surface, and burn it to generate the steam.

In various places around the Rift Valley there is a high heat flow, blatant in its jets of steam. On my first visit to Kenya, when pausing on the magnificent scenic road that looks down on Lake Naivasha, I had been intrigued by the considerable clouds of steam emerging from rocky hills to the lake's south-west. Years later I stayed with friends lucky enough to have a home by that delightful lake, and we often used to fly a hot-air balloon from their hippo-cropped lawn. Generally we flew north, past Crescent Island, over papyrus and then thorn trees to land near a portion of the famous railway line built from Mombasa to Victoria. Occasionally we travelled south, receiving a stunning view of that steam bursting forth from down below. It was not a volcano but, in similar fashion, gave warning of the great forces beneath the superficial crust. We did not land there, the rocks being too unwelcoming, but we never forgot that awesome sight – a piece of the earth blowing off steam as any kettle might do, save that no one was in charge of this gigantic kettle.

Later, in the early 1970s, the Kenyan government decided to investigate this likely source of power. Six experimental bore-holes were sunk, the deepest reaching over a kilometre and a half below the surface. It was all very well looking at the steam, as we had from our altitude, but its precise quantity had to be measured, its corrosive qualities had to be understood, and its capabilities calculated. Would sufficient electricity be generated to make the drilling and its transmission a worthwhile enterprise? Would the steam last when being tapped near its point of origin?

The trials were satisfactory, and in the late 1970s generation equipment was installed at the site now known as Olkaria. Currently this single place is producing 16 per cent of Kenya's power. The depths from which the steam is emerging are considerable, varying from 800 metres to nearly 2200 metres. Some of the exploratory wells sunk since production began are deeper still. The area being tapped has a temperature of 250-300° Centigrade. This is hotter than water's normal boiling point, but it is still liquid at such depths because the pressure there is many times greater than at the surface. The pressure decreases as the water ascends, turning

214

much of the liquid into steam, and the mix at the surface is about 20–30 per cent steam and 70–80 per cent water. These are separated, the steam being despatched to drive the turbines, as in any normal power station, and the water is wasted. It is allowed to seep into the ground and no one knows whether, eventually, it emerges as more steam or travels somewhere else.

A few geothermal stations in other countries employ a system of re-injection, sending the hot water down to boost the steam supply, but this is not yet happening at Olkaria. Unfortunately, in a country and an area perennially short of water, this hot excess cannot be cooled and used for irrigation. Even at depth it is highly charged with minerals and, when its steam has been diverted, the remainder is even more so. Its principal constituents are silicates and chlorides, with more unfortunate extras, such as arsenic, lithium and boron.

One disappointment with much of industry is that the casual observer cannot see what is actually happening. The action is generally contained and, however much registered on dials, is nonetheless invisible. This is certainly true of an oil well, a gas well, or even a steam well. It would be careless, and inefficient, to let such items blow off in a conspicuous manner. However, with the experimental steam bore-holes, drilled to discover how much energy is actually available, the outflow is all too visible. It is also deafening and, downwind, dampening. I found it horrendously impressive, never having seen or heard or felt such a quantity of steam. It dwarfed any form of steam-engine previously encountered, pouring 30 tonnes an hour from an expansion tube over a metre in diameter, substantially wider than the original 23-centimetre bore-hole. The vent was turning a bright blue sky into a cloudy day as this different kind of cloud roared forth and then drifted off downwind.

The well was over 1800 metres deep and the engineers thought that the actual steam/water mix was coming from a depth nearer 1400 metres. Bore-holes are usually tested for a couple of months, and at 30 tonnes of steam an hour, this means the loss of 44000 tonnes into the atmosphere without the generation of a single watt. It seems a waste, but it is useless to install generating equipment on top of an outlet before checking whether it can maintain a reasonable supply without the pressure dropping significantly. The pressure will inevitably diminish in time, and it is important to assess whether the life of the well is likely to justify the cost involved in making use of its libation of energy. The well that was deafening my ears, clouding the day, spoiling my notebook, and giving violent proof of pressures existing down below, was thought to be satisfactory. It was estimated that it would generate 4 megawatts of electricity, enough to light 66 000 bulbs each of 60 watts and, therefore, not trivial. There is the cost of the bore-hole, the electrical equipment and the transmission lines, but the basic energy to drive the turbines is all quite free, one blessing of living by a Rift Valley.

To achieve a more distant view of the Olkaria site, and to leave that ear-numbing roar of steam, I climbed a nearby lava flow. It seemed entirely apposite, scrambling up one aspect of the Rift system to observe another. An occasional shrub was growing from the rock where some earth had gathered to give it purchase; but, for the main part, the lava looked much as it had done on that fiery day, perhaps a few hundred years ago, when it emerged from the depths to solidify upon the land. The twists and curls it had fashioned then, as portions cooled and others flowed above them, are no less plain today. The contortions are most conspicuous, as if in death throes from the agony of having to become solid so soon after being released. Embedded in them, much like the fillings of some multi-layered cake, were folds of obsidian, the black and natural glass which was such a boon to many of our ancestors. Flint was highly prized wherever it occurred for its sharpness and strength. Obsidian could be flaked in similar fashion and, for stone-age people, was of considerable benefit as a cutter, a scraper, an arrow-head or spear tip, or even as an ornament. For modern people, supposedly with some sense to them, its cutting edge proves itself all too readily if attempts are made with fingers to extract it from the rock. Time and again in the past that sharpness must have been a virtue, even if stone-age fingers also occasionally suffered when gouging it from the rock. As obsidian is a form of rhyolite, rich in silica and more viscous than most lavas, the contortions were readily explained. That particular flow was not able to travel far before its cooling stiffness put an end to it. The hill up which I scrambled (and cut myself rather more than once) was indeed no more than a stubby promontory.

From its summit I could see not only its twisted length and breadth, like a foaming, breaking, fossilised piece of sea, but all the Olkaria site. There were few of the trappings of a normal power station, such as cooling towers or a passing river to fulfil the same role. Instead, there were many scattered well-heads over the little hills and valleys of that uneven ground. From each such point was a small wisp of steam, with shiny pipes running as interconnections between them and various buildings. It all looked most relaxed, and extremely convenient. People have always taken what has been on offer, and according to their abilities. First it was obsidian and wild beasts. Now it is steam at 250-300° Centigrade from 900 metres or more below the surface. There is not much in common between the silica glass and the hot steam, save that they are both by-products of a rift.

There are also similarities between steam gushing from the Earth and water pouring down a mountainside. The steam has to be discovered and the mountain water has to be dammed. Although the basic energy may cost nothing in either case, there are inevitable expenses before electricity can hum along the wires. There are many lakes within the Rift system but not the flow that can be tapped for electricity. The greatest hydroelectric station in the area, creating 150 megawatts, is at Jinja where Lake Victoria overflows to start the White Nile on its journey. However, that lake is not a true Rift lake. Water has concentrated there because some slightly lower-lying land proved convenient for a couple of major rivers. The two strings of Rift lakes are often deep, but most of them have no surface outlets, and therefore no rush of water to be tapped by turbines. As a further impediment to hydroelectricity the rainfall in much of the Rift's area is either erratic or sparse or both. It is therefore fair, if that word has any validity when discussing natural blessings, that a region without much hydroelectric potential has a compensatory source of free power.

Unlike Magadi's soda, geothermy is not a sustainable resource. The pressures being tapped at Olkaria are dropping. The heat still exists, but not the water supply to bring it to the surface. The plans for re-injecting water involve extra expense and more complexity. One problem is caused by separating the steam from the water, when the steam is taken to drive the turbines. The water, already contaminated with minerals at its point of origin, becomes richer still after the separation. To make use of such a super-saturated solution is probably to suffer depositions of, for example, silica, and the liquid will become yet more saturated each time that it is used. A better plan would be to use fresh water for the re-injection, but that would be demanding upon local water, almost always in short supply.

Other sites are also being investigated, notably to the north of Lake Baringo. Unfortunately, as with hydroelectric power, such free energy

217

is not necessarily located where it would be most welcome. Olkaria is not too far from developed areas, but the region between lakes Baringo and Turkana is a sparsely populated zone. It is magnificently scenic, an excellent portion of the Rift Valley which excited John Gregory on his epic journey of geological exploration; but it is a difficult place to live. A high heat flow, suitable for geothermy, occurs only where it chooses to occur. The mineral deposits, and other natural blessings, were placed without any reference to the need for growing things, and therefore occur anywhere. Potentially interesting geothermal sites obey this universally unhelpful rule.

Little is known about the origin of the Olkaria water, which becomes such useful steam. Whatever its source, a very great deal of liquid is involved, and the Olkaria station will process some 13 million tonnes of water in a 10-year span. Naivasha's level is falling and, since Olkaria is only a few kilometres from the south-western edge of the lake, the suggestion has inevitably arisen that Naivasha is a source, if not *the* source. The lake is large, but the loss of 13 million tonnes would lower its level by about 46 centimetres. However the steam geologists believe that Olkaria water comes from the hilly country to the west. Naivasha is one water table but, so they say, the bore-holes pass down through two or three more aquifers before reaching the geothermal level. Natural radioactive isotopes, present to different degrees in every body of water, also suggest that the steam originates in the westerly hills.

If the steam take-off is the culprit it is also draining lakes Elmenteita and Nakuru, Naivasha's two nearest neighbours lying to the north-west. They are extremely low, to the detriment of the flamingoes at Nakuru and to the general sight of Elmenteita, which is fast becoming little more than a mud-plain. However, Olkaria cannot be held responsible for the general lowering of the Rift lakes which has been occurring in the late 1980s. It has been suggested that lower rainfall in Ethiopia is causing the drop in Turkana, and the shrinking Lake Tanganyika is certainly receiving less inflow than usual via the Ruzizi from Lake Kivu. Possibly the cause is as general as the lowering. Maybe there is some subterranean heaving, akin to the creation of the Rift itself, that is shrinking all the levels in some unified manner. Perhaps geothermy is to blame, in part, for Naivasha's lowering but, with so many other culprits, and with doubt about the source of all that superheated steam, no one will switch off 16 per cent of Kenya's power on such slender evidence.

Just as geothermy is one likely by-product of a Rift Valley system, so are reserves of oil. Large quantities have already been discovered in areas of a similar geological structure, such as the Gulf of Suez and the North Sea. These too are graben basins and, although the Great Rift of Africa may look strikingly dissimilar, the geological parallels make the oilmen optimistic. They know there has to be not only a source of oil but traps for the gathering of that oil. Major faulting can provide suitable traps and a rift system beset with faults is therefore highly suitable as a likely spot for capturing black gold.

During the early 1980s attention began to be focused on the Rift as an area worthy of extensive exploration. The areas being particularly investigated are on both sides of Lake Turkana, the northern Kenyan border area with Somalia, and Lake Tanganyika. Advantages as well as problems exist in remote areas, with a major virtue being relatively easy access. Only nature has to be contended with, and none of the snags associated with a developed area. The most taxing region under examination is Lake Tanganyika, the Rift's deepest lake. It is less turbulent than, for example, the North Sea, but its depth is inconvenient.

In early 1988 reports surfaced that 'basins with oil potential' were being discovered in northern Kenya. The companies involved have been spending ten of millions of dollars, and they presumably do not use such quantities without hope of recovery. The actual drilling is much more expensive than the surface prospecting, but drilling has already begun optimistically. By no means does every drill discover oil, or even positive signs that oil exists. Once oil is found, quite a few bore-holes must be sunk before anyone knows whether there is sufficient for a pipeline or whether trucking the oil will be a more sensible alternative.

Financially, the discovery of oil may be welcome, or even a bonanza, but there will be negative aspects. Oil exploration surveys have been carried out on lakes Turkana, Albert, Victoria, Tanganyika and Malawi. Tanganyika would be particularly vulnerable to the disturbance any exploitation might entail. This lake's fauna are probably the most scientifically valuable of any lake in the world. Its pelagic fishery, to quote a recent paper on the special problems of the African great lakes, 'appears potentially to be one of the largest in Africa and a natural food resource of immense value . . . When fully exploited, the catch from the lake may provide a large part of the protein requirements of several million people in the lakeshore countries'. Its 19,200 cubic kilometres form the largest single reservoir of freshwater on the continent. The mixing of its water layers would, in itself, be a disaster, for three-quarters of its volume contains no oxygen and would certainly be lethal to all the life existing in the surface layers. The lake does receive an inflow from two rivers, but hundreds of years would be necessary for the lake to be refilled. Any change to the lake's waters, now so acceptable to a unique range of animals, could not be put right for an unthinkable length of time. The oil men must realise, if exploration does become exploitation, that they will be trespassing upon a most delicately balanced region.

The difficulties caused by drilling for permanent water have been acute, and no one argues that oil will be less painful in the changes it creates. Of course the drillers can be maintained, and all their worldly needs imported, but will the local people – the Rendille, Boran, Samburu, Turkana – be able to persevere with ancient ways cheek by jowl with modern times? Will they want to? Some day in the future every reserve of oil that is discovered will run dry, and the area may have to revert to its former ways. To look down upon the Rift is to see great beauty, staggering contrasts and delicately balanced environments. The Rift offers an extraordinary range of habitats to animals and humans alike. Now its minerals, formed by the powerful forces still at work in the region, are starting to provide much-needed wealth. As someone privileged to explore the Rift, and who has seen it from all manner of angles (even if a few did terrify at the time), I hope a way can be found to exploit these natural blessings without upsetting the fine balance of this unique area – the greatest Rift system on any continent.

PICTURE CREDITS

—

INDEX

4

916.76
S   Smith, Anthony
       Th Great Rift.

916.76
S   Smith, Anthony
       The Great Rift.

4 82

89-1468

9-21 Lillian Moss

R 10-5

APR 24   K. Gordon-Bey